A SHORT COURSE
IN
GEOMETRY

PRENTICE HALL
Upper Saddle River, New Jersey 07458

A SHORT COURSE IN GEOMETRY

PATRICIA JUELG
Austin Community College

San Francisco, California

Divisions of Macmillan, Inc.

Toronto

Library of Congress Cataloging-in-Publication Data

Juelg, Patricia.
   A short course in geometry/Patricia Juelg.
      p.   cm.
   ISBN 0-02-361471-4
   1. Geometry.   I. Title.
QA453.J82   1990
516—dc20   90-31093
                                                                      CIP

On the cover: A detail from a painting by Terence La Noue, 1988–89; mixed media on canvas. In his work, La Noue combines abstraction with motifs taken from many cultures, contrasting geometric and organic shapes on his large, tapestry-like paintings that hang freely from the wall. His work is in several public and private collections, including the Whitney Museum of American Art and the Guggenheim, New York City, and the Musée de Strasbourg, France. La Noue is represented by Dorothy Goldeen Gallery in Santa Monica, California, and by André Emmerich Gallery in New York.

Photo credits: Photos on pages 1, 49, and 189 by Rod Dresser; photo on page 85 by Bill Lemke/Third Coast Stock Photos; photo on page 157 by A. Devaney, Inc.; photo on page 245 by Amwest.

© 1991 by Prentice-Hall, Inc.
A Pearson Education Company
Upper Saddle River, NJ 07458

Printed in the United States of America

ISBN 0-02-361471-4

Prentice-Hall International (UK) Limited,London
Prentice-Hall of Australia Pty. Limited, Sydney
Prentice-Hall Canada Inc., Toronto
Prentice-Hall Hispanoamericana, S.A., Mexico
Prentice-Hall of India Private Limited, New Delhi
Prentice-Hall of Japan, Inc., Tokyo
Pearson Education Asia Pte. Ltd., Singapore
Editora Prentice-Hall do Brasil, Ltda., Rio de Janeiro

CONTENTS

                         5.1   AREA   190
                         5.2   SURFACE AREA   217
                         5.3   VOLUME OF SPACE FIGURES   229
                               REVIEW EXERCISES   241
                               PRACTICE TEST   243

CHAPTER SIX   **CONGRUENT AND SIMILAR TRIANGLES;**
                         CONSTRUCTIONS; AND COORDINATE GEOMETRY   **245**

                         6.1   CONGRUENT TRIANGLES   246
                         6.2   SIMILAR TRIANGLES   255
                         6.3   CONSTRUCTIONS   265
                         6.4   COORDINATE GEOMETRY   281
                               REVIEW EXERCISES   291
                               PRACTICE TEST   295

                         ANSWERS   297
                         INDEX   305

PREFACE

AUDIENCE

This text is written for use in colleges and assumes that a student has completed an introductory course in algebra. It can be used effectively as a one-semester-hour course in geometry or as a supplement for a course that requires a short segment on geometry. This text gives students the essential elements of geometry needed to be successful in any course requiring a basic geometry prerequisite.

CONTENT

The text contains the fundamental topics found in a three-semester-hour geometry course, but the topics are condensed and the material judged nonessential has been deemphasized. For an even shorter course Sections 1.1, 1.2, 1.3, 1.4, Chapter 2, and Sections 6.3 and 6.4 may be omitted.

CHAPTER ORGANIZATION

Objectives Every section begins with a list of objectives that are covered in the section. These objectives are flagged and numbered.

Quick Checks The text is organized so that after each skill is developed there is a Quick Check exercise. The Quick Checks may be used by the instructor as lecture examples or by the student to check his or her understanding of the material. Answers to the Quick Checks are given at the bottom of the page on which the Quick Check appears.

Section Exercises At the end of each section there is a set of exercises that is designed to test the student's mastery of the concepts developed in the section. Answers for the odd-numbered exercises appear in the back of the book. Most exercise sets contain A and B exercises. The B exercises ask students to stretch the concepts they have learned and often involve more complicated problem solving. Furthermore, these exercises may be effectively used as bonus-point problems.

Chapter Review Exercises At the end of each chapter is a Review Exercise that is a comprehensive review of the material discussed in the chapter. These reviews are designed for coverage in a 50-minute review lecture. Answers for the chapter review exercises appear in the back of the book.

Chapter Practice Tests Following each Review Exercise is a Practice Test. These tests are not designed to be comprehensive; instead, they may be considered as condensed reviews. Answers for the chapter practice tests appear in the back of the book.

ANCILLARIES

Instructor's Manual   The instructor's manual contains the following:

       Chapter-by-chapter notes to the instructor

       Section-by-section lists of topics contained in each section

       Cumulative reviews for every three chapters

       Multiple-choice tests for each chapter

       A diagnostic pretest for the course

       Two versions of a final exam

       Transparency masters

       Complete solutions for the even-numbered exercises

Student Self-Help Package   The student self-help package contains the following:

       A student manual containing hints and the complete, annotated solutions for odd-numbered exercises

       Videotapes prepared by the author, which present motivations for topics, essential core concepts, applications, and worked-out examples

Computer-Generated Tests   The computer-generated tests, prepared for the Apple IIE and IBM computers, provide multiple-choice and open-ended tests for each chapter. The numbers and expressions in the tests are randomly generated, allowing as many iterations of a test as desired by the instructor. Answers for each test are computer-generated.

ACKNOWLEDGMENTS

The author would like to thank the following people for helping to make this text possible: Donald E. Dellen, Phyllis Niklas, Lilian Brady, Beth Anderson, Diane Harmon, Kylene Norman, and John Bailey.

A SHORT COURSE
IN
GEOMETRY

CHAPTER ONE

LOGIC
AND SETS

1.1 THE LOGICAL STRUCTURE OF MATHEMATICS

OBJECTIVES

1 ▶ Explain the logical structure of mathematics.

2 ▶ Explain what is meant by an undefined term.

3 ▶ State the requirements for a mathematical definition.

4 ▶ State the difference between a property (or axiom) and a rule (or theorem).

1 ▶ ## THE LOGICAL STRUCTURE

Mathematics is a logical system of laws and definitions dealing primarily with collections of numbers, points, and geometrical figures. Figure 1.1 shows a breakdown of the logical structure of mathematics. Structurally, mathematics begins with a collection of undefined terms, followed first by a set of mathematical definitions and then by a collection of properties or axioms. Based on the properties a collection of rules or theorems can then be developed. Once the undefined terms, definitions, properties, and rules have been established, the structure branches into problems, drill and practice, applications, and ideas for new rules. Once this mathematical base is established, the concepts may be applied to many fields. To be successful in the study of mathematics you must understand its logical structure, be able to operate within its established boundaries, and apply the tools of mathematics to everyday problems.

FIGURE 1.1
The logical structure of mathematics

▶**2** UNDEFINED TERMS

At first, the idea of an undefined term may seem strange. You have been led to believe that every word has a definition because you can look it up in the dictionary. Dictionaries use circular definitions; that is, they assume that somewhere in the circle of definitions you will intuitively know a word and will therefore know all of the words in the circle. For example, if you look up the word "hop" in a dictionary, it means "jump"; if you look up "jump," it means "spring"; if you look up "spring," it means "leap"; and if you look up "leap," it means "hop" (Figure 1.2). The dictionary assumes that you will find a word for which you intuitively know the meaning as you proceed around this circle of definitions. In mathematics, words that you are expected to know intuitively are called **undefined terms**.

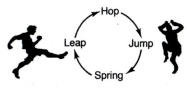

FIGURE 1.2
Circular definition

Undefined terms in mathematics are designated at the beginning so that you are aware of the words you are expected to know intuitively. The undefined terms that we use in this course are "point," "line," "plane," "space," "set," and "number." That is, you are expected to know intuitively the meanings for these terms. We do not try to define a point, a line, a plane, space, a set, or a number; instead, we discuss their characteristics and teach you how to operate with points, lines, planes, space, sets, and numbers.

▶**3** MATHEMATICAL DEFINITIONS

In mathematics a set or collection of well-defined terms is called a **mathematical definition**. For a statement to be considered a mathematical definition it must meet the following requirements.

REQUIREMENTS FOR A MATHEMATICAL DEFINITION

1. The word or phrase being defined must be placed in a set or collection.
2. The definition must tell how the word or phrase can be distinguished from all other members of the set.

If a statement does not meet the preceding requirements, it is not a mathematical definition.

EXAMPLE 1

Which of the following statements are mathematical definitions?

1. A "bear" is a very large animal.
2. A "puppy" is a dog less than one year old.
3. A "joke" is a story that makes you laugh.
4. "Mathematics" is a science.
5. A "rose" is a flower.
6. A "rubber ball" is a spherically shaped object made of rubber.

Solution

1. The statement is not a mathematical definition because it does not tell how a "bear" can be distinguished from other animals in the set.
2. The statement is a mathematical definition because the word "puppy" is placed in the set of all dogs and then distinguished from other dogs in the set by saying that a "puppy" must be less than one year old.
3. The statement is a mathematical definition because the word "joke" is placed in the set of stories and is distinguished from other stories by saying that it is a story that makes you laugh.
4. The statement is not a mathematical definition because it does not distinguish "mathematics" from the other sciences.

5. The statement is not a mathematical definition because it does not tell how a "rose" is distinguished from all other flowers in the set of flowers.
6. The statement is a mathematical definition because it places "rubber ball" in the set of all spherically shaped objects and then tells how it can be distinguished from other spherically shaped objects in the set—it is rubber.   ◀

QUICK CHECK

Which of the following are mathematical definitions?

1. A "teenager" is a person who is 13, 14, 15, 16, 17, 18, or 19 years old.

2. A "raccoon" is a four-legged animal.

3. A mixture of lemon, water, and sugar is called "lemonade."

4. A "quarter" is a U.S. coin worth 25 cents.

5. "Pink" is a beautiful color.

▶ PROPERTIES AND RULES

In arithmetic and algebra there are 11 fundamental rules called the **Field Properties** or Field Axioms. These properties are taken to be true without argument and form the foundation for the structure of arithmetic and algebra. For example, the Commutative Property for Addition states that the order of addition may be reversed; that is, $3 + 7 = 7 + 3$. The study of geometry is governed by a set of properties relating to various sets of points.

Based on the Field Properties and the properties of geometry, a set of **rules** or theorems can be established. For example, a rule in arithmetic states that $-1 \cdot 3 = -3$. These rules are justified or proven true by definitions, previously proven rules, and properties. That is, if the properties are accepted, the rules are statements that must logically follow.

ANSWERS

Numbers 1, 3, and 4 are mathematical definitions.

A

ANSWERS

Answer each of the following true or false.

1. Undefined terms have no meaning.

2. You are expected to know the meaning of undefined terms intuitively.

3. Properties and rules are both rules.

4. Properties and rules are the same.

5. The statement "A sad movie is a movie that makes me cry" meets the requirements for a mathematical definition.

6. A mathematical definition places the word or phrase being defined in a set and tells how it can be distinguished from all other members of the set.

Determine whether each of the following is a mathematical definition. If it is not, explain why.

7. An orange is a fruit.

8. A banana is a tropical fruit.

9. A horse is a four-legged animal.

10. A dog is a domestic animal.

11. The numbers 0, 1, 2, 3, 4, 5, 6, 7, 8, and 9 are called digits.

12. The numbers 1, 2, 3, 4, 5, . . . are called natural numbers.

13. A number that is exactly divisible by 2 is called an even number.

14. A number that is not exactly divisible by 2 is called an odd number.

15. A number that is exactly divisible by 5 is a multiple of 5.

16. A number that is exactly divisible by 6 is a multiple of 6.

1. _____

2. _____

3. _____

4. _____

5. _____

6. _____

7. _____

8. _____

9. _____

10. _____

11. _____

12. _____

13. _____

14. _____

15. _____

16. _____

1.2 CONDITIONAL AND BICONDITIONAL STATEMENTS

OBJECTIVES

1▶ Find the converse of a statement.

2▶ Identify a conditional statement, find its converse, and determine whether the statement or its converse is true.

3▶ Recognize biconditional statements and state what must be true if a biconditional statement is true.

1▶ STATEMENTS AND CONVERSES

A statement is simply a declarative sentence that is either true or false. A **mathematical statement** is a statement pertaining to some aspect of mathematics and generally involves mathematical relationships, symbols, and operations. The **converse** of a statement is a new statement that is made by interchanging the clauses or phrases that appear before and after the verb.

EXAMPLE 2   Find the converse for each statement.

1. Statement: A rose is a flower.
2. Statement: A square is a rectangle.

Solution   1. The converse is formed by interchanging "a rose" and "a flower." Converse: A flower is a rose.
2. To form the converse, interchange "a square" and "a rectangle." Converse: A rectangle is a square.   ◀

A statement may be true and its converse false, or the statement may be false and its converse true, or they may both be true or both be false. However, when a statement is a mathematical definition, both the statement and the converse are true.

> Both the statement and the converse of a mathematical definition must be true.

For example, consider the following statement and its converse.

Statement: *The numbers 0, 1, 2, 3, 4, 5, 6, 7, 8, and 9 are the digits.*

Converse: *The digits are the numbers 0, 1, 2, 3, 4, 5, 6, 7, 8, and 9.*

Since the statement is a mathematical definition, both the statement and its converse are true.

QUICK CHECK

Find the converse for each statement.

1. Statement: A rubber ball is a spherically shaped object made of rubber.

2. Statement: A four-sided polygon is a rectangle.

3. Statement: A pumpkin is an oyster.

▶ 2 CONDITIONAL STATEMENTS

Sentences that contain an "if" clause followed by a "then" clause are called **conditional statements**. In a conditional statement the "if" clause tells what is assumed true; the "then" clause states what must logically follow when the "if" clause is true.

CAUTION ▶ If a conditional statement is true, its converse is not necessarily true. When a mathematical statement is written in conditional form, it does not necessarily follow that its converse is also true.

EXAMPLE 3 Each of the following is a true conditional statement. Find the converse for each statement and determine whether the converse is true.

1. If it is a square, then it is a rectangle.
2. If it is a rose, then it is a flower.
3. If it is a rubber ball, then it is a spherically shaped object made of rubber.

Solution 1. Converse: If it is a rectangle, then it is a square. The converse is false because all rectangles are not squares—a rectangle 3 feet by 4 feet is not a square.
2. Converse: If it is a flower, then it is a rose. The converse is false because all flowers are not roses. For example, a daisy is a flower but it is not a rose.
3. Converse: If it is a spherically shaped rubber object, then it is a rubber ball. The converse is true. ◀

ANSWERS

1. A spherically shaped object made of rubber is a rubber ball.
2. A rectangle is a four-sided polygon.
3. An oyster is a pumpkin.

QUICK CHECK

Decide whether each statement is true, write its converse, and determine whether the converse is true.

1. If there are no clouds, then I can see the sun.

2. If I water my potted plant properly, then it will grow.

▶ **3** BICONDITIONAL STATEMENTS

A sentence formed by joining a conditional statement and its converse with the word "and" is called a **biconditional statement.**

> *If it is a triangle, then it is a three-sided polygon, and if it is a three-sided polygon, then it is a triangle.*

Biconditional statements may be shortened by using the phrase "if and only if," which signals that the statement is biconditional. Using this phrase the preceding biconditional statement may be shortened to

> *It is a triangle if and only if it is a three-sided polygon.*

The phrase, "if and only if" is frequently abbreviated "iff," so our example becomes

> *It is a triangle iff it is a three-sided polygon.*

For a biconditional statement to be true, both the conditional statement and its converse must be true. When a mathematical definition or rule is stated in biconditional form, both the statement and its converse are true.

> A mathematical definition may be written in the "if and only if" form because both its statement and converse must be true.

EXAMPLE 4 For each of the following statements, (a) decide whether each statement is true, (b) write its converse, (c) determine whether the converse is true, (d) write the biconditional statement, and (e) state whether the biconditional statement is true.

1. If the digit in the units place of a whole number is a 5 or a 0, then the number can be divided by 5 and it will have a zero remainder.
2. If the colors yellow and blue are mixed, the resulting color will be green.
3. If I study history for two hours every night, then I will pass.

ANSWERS

1. The statement is not necessarily true because if it were night and there were no clouds, you would not be able to see the sun. Converse: If I can see the sun, then there are no clouds. The converse is not true because you can see the sun on a partly cloudy day.
2. The statement is not necessarily true because it takes more than just proper watering to make a plant grow. Converse: If my potted plant grows, then I have watered it properly. The converse is true because if the potted plant grows, it must have been watered properly.

Solution   1. a. The statement is true.
              b. Converse: If a number can be divided by 5 with a zero remainder, then
                 the digit in its units place is a 5 or a 0.
              c. The converse is true.
              d. Biconditional statement: A number can be divided by 5 with a zero re-
                 mainder iff its units digit is a 5 or a 0.
              e. The biconditional statement is true because both the statement and its con-
                 verse are true.
           2. a. The statement is true.
              b. Converse: If the color is green, then the colors yellow and blue have been
                 mixed.
              c. The converse is true.
              d. Biconditional statement: The color is green iff the colors yellow and blue
                 have been mixed.
              e. The biconditional statement is true because both the statement and its
                 converse are true.
           3. a. The statement is not necessarily true; therefore it is false.
              b. Converse: If I pass history, then I have studied for two hours every night.
              c. The converse is not necessarily true.
              d. Biconditional statement: I will pass history if and only if I study history for
                 two hours every night.
              e. The biconditional statement is not true because the validity of the state-
                 ment and its converse is questionable.   ◄

EXERCISE 1.2

A

ANSWERS

Answer each of the following true or false.

1. Mathematical definitions should be written as conditional statements.

    1. _____

2. Mathematical definitions can and should be written in the "iff" form.

    2. _____

3. If a statement and its converse are both true, then the corresponding biconditional statement is true.

    3. _____

4. If a conditional statement is true, then its converse is also true.

    4. _____

5. All conditional statements are true.

    5. _____

6. Mathematical statements are sentences pertaining to mathematics.

    6. _____

7. When the "if" clause of a conditional statement is true, the entire statement is true whether or not the "then" clause is true.

    7. _____

8. "If and only if" is abbreviated "iff."

    8. _____

Write each of the following as a conditional statement.

9. A robin is a bird.

    9. _____

10. A trout is a fish.

    10. _____

11. A 300-mile car trip makes me tired.

    11. _____

12. Studying improves my grades.

    12. _____

13. Eating strawberries gives me hives.

    13. _____

Write the converse of each statement.

14. If I get eight hours of sleep, then I feel energetic.

    14. _____

15. If the temperature is over 90 degrees, then it is hot.

    15. _____

16. If it is a circle, then it has a center.

    16. _____

17. If it is a pyramid, then it comes to a point.

    17. _____

18. If an even whole number is divided by 2, then the result will have no remainder.

    18. _____

Write each of the following in "if and only if" form.

19. A tree is a plant.

    19. _____

20. A square is a four-sided polygon with sides of equal length.

    20. _____

21. The digits are the numbers 0, 1, 2, 3, 4, 5, 6, 7, 8, or 9.

    21. _____

22. Whole numbers exactly divisible by 10 end in the digit 0.

    22. _____

23. Eskimos live at the North Pole.

    23. _____

B

Define each term and write the definition as a biconditional statement.

24. A number divisible by 4   24. _____

25. A number divisible by 11   25. _____

26. A basketball   26. _____

27. A football   27. _____

1.3 "AND" AND "OR" STATEMENTS

OBJECTIVES

▶1 Given a true "and" statement, determine the validity of a conclusion.

▶2 Given a true "or" statement, determine the validity of a conclusion.

▶1 "AND" STATEMENTS

A **conjunction** is a statement formed by connecting two statements with the word "**and.**" For a conjunction to be true, both statements joined by the "and" must be true. For example, consider the following conjunction:

The car Margie buys must be red, and it must have only two doors.

For the entire statement to be true, both conditions must be satisfied. That is, the car must be red and it must have exactly two doors or Margie will not buy it. She would not buy a red four-door car, or a green two-door car, or a blue four-door car.

Consider this conjunction:

It is a number, and it is divisible by 2 with no remainder.

Three is a number but it is not divisible by 2, so 3 is not a number described by this conjunction. On the other hand, 46 is a number and is divisible by 2. Therefore, 46 is one of the numbers described by this conjunction.

EXAMPLE 5   Assume that each statement is true and determine whether the conclusion is true or false.

1. Statement: The hat Jerry buys must be black, and it must be a baseball cap. Conclusion: Jerry will buy a black Stetson.
2. Statement: Abel said, "The person I hire for this job must be able to type 75 words per minute and must take dictation at the rate of 90 words per minute. Conclusion: Sammy can type 80 words per minute and take dictation at the rate of 95 words per minute; therefore, Abel will consider hiring Sammy.

Solution   1. The conclusion is false because, although the hat is black, it is not a baseball cap.
2. The conclusion is true because Sammy meets both the requirements for the job.   ◀

> ### QUICK CHECK
>
> *Assume that the following statement is true and determine whether the conclusions are true or false.*
>
> *Ben is going to buy a jacket that is blue and a windbreaker.*
>
> *Conclusions:*
>
> 1. Ben will buy a green windbreaker.
> 2. Ben will buy a blue wool blazer.
> 3. Ben will buy only a blue windbreaker.

2 ▶ "OR" STATEMENTS

When two statements are joined by the word "**or**," the sentence formed is called a **disjunction**. An "or" statement or disjunction is true when at least one of the statements joined by the "or" is true. For example, when we say "Jim will buy a blue car or a four-door car," we mean that Jim will buy a blue car, a four-door car, or a blue four-door car. Therefore, only one of the statements connected by the "or" must be true for the disjunction to be true.

Consider the following disjunction:

It is a number or it is divisible by 2.

The number 3 would be one of the numbers described by the statement because it is a number. The number 4 would also qualify as one of the numbers described by the statement because it is a number and is also divisible by 2.

EXAMPLE 6

Assume that each statement is true and determine whether the conclusion is true or false.

1. Statement: Rose said that the dress she buys must be red or be made of wool.
   Conclusion: Rose will buy a red cotton dress.
2. Statement: It must be a digit or it must be the number 15.
   Conclusion: It could be 11.

Solution

1. The conclusion is true because an "or" statement only requires that at least one of its conditions be true. The dress she selected is red.
2. The conclusion is false because 11 is neither a digit nor 15. ◀

ANSWERS

1. False 2. False 3. True

QUICK CHECK

Assume that the following statement is true and determine whether the conclusions are true or false.

Joey is going to buy a jacket that is blue or that is a windbreaker.

Conclusions:

1. Joey will buy a green windbreaker.

2. Joey will buy a blue wool blazer.

3. Joey will buy a blue windbreaker.

ANSWERS

1. True   2. True   3. True

◄ E X E R C I S E 1 . 3

A

ANSWERS

Assume that each statement is true and determine whether the conclusions are true or false.

Statement: Sarah is going to buy a red dress or a blue dress.

1. Conclusion: Sarah will not buy a green dress.

2. Conclusion: Sarah will buy a red dress.

1. _____

2. _____

Statement: It is an even whole number between 0 and 10 (not including 0 or 10) or it is a whole number larger than 7.

3. Conclusion: It could be 3.

4. Conclusion: It could be 4.

5. Conclusion: It could be 11.

6. Conclusion: It could be 7.

7. Conclusion: It could be 8.

3. _____

4. _____

5. _____

6. _____

7. _____

Statement: We are having steak for lunch, and we are also having fried potatoes.

8. Conclusion: Lunch will consist of steak only.

9. Conclusion: Lunch will consist of potatoes only.

8. _____

9. _____

Statement: It is an even whole number, and it is less than 20.

10. Conclusion: It could be 5.

11. Conclusion: It could be 6.

12. Conclusion: It could be 22.

13. Conclusion: It could be 19.

10. _____

11. _____

12. _____

13. _____

B

Answer each of the following true or false.

14. For an "or" statement to be true, both clauses must be true.

15. For an "and" statement to be true, both clauses must be true.

14. _____

15. _____

1.4 LOGIC

OBJECTIVES

▶1 Given several true statements, use deductive reasoning to determine the validity of a conclusion.

▶2 Use inductive reasoning to draw valid conclusions.

▶1 DEDUCTIVE REASONING

Deductive reasoning is a kind of logic in which a statement or group of statements is assumed or known to be true and conclusions are then drawn from the statements. Much of mathematics is based on deductive reasoning. In math the assumed facts are the Field Properties, the properties of geometry, and many mathematical rules and procedures are derived directly from these properties by using deductive reasoning.

Science, on the other hand, is primarily based on a second kind of logic known as inductive reasoning. **Inductive reasoning** is a kind of logic in which conclusions are drawn on the basis of observations. When we reason inductively we are looking for patterns. Based on our understanding of the pattern, we predict the next occurrence in the pattern. For example, a biologist who observes the behavior of a particular colony of apes predicts the behavior of all apes of the same kind based on his observations.

Deductive reasoning is the application of a general statement that is known or assumed to be true in a particular situation. On the other hand, inductive reasoning is an attempt to use a number of specific observations to make a generalized statement of truth based on the observations. In short, deductive reasoning goes from general to specific, whereas inductive reasoning goes from specific to general.

Frequently we need to piece together a number of known facts to solve a problem. Drawing a sketch representing the known information often helps to determine the solution. We use this approach in the following example.

EXAMPLE 7

John's house is behind Nita's house.

Pat's house is behind Marge's house.

Pat lives next door to Nita.

John lives next door to Sam.

None of these people lives on a corner.

1. Do John and Pat live on the same street?
2. Does Pat live next door to Sam?
3. Does Sam live behind Pat?
4. Does John live next door to Marge?

Solution

To answer the questions, draw a sketch showing the possible arrangements of the houses.

The last given statement tells us that the only way any of these people could live next door to two of the others is to live between them on the same street.

The first given statement (that John lives behind Nita) can be represented by the following sketch.

The second given statement has nothing to do with the people named in the first statement so let's skip it temporarily and go to the third statement. We are told that Pat lives next door to Nita, but we don't know which side of Nita she lives on. Thus either of the sketches shown below could represent the situation as we know it so far.

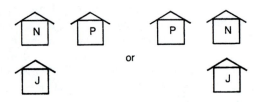

Now we can use the second statement. Since Marge's house is behind Pat's house, one of the following sketches represents the situation.

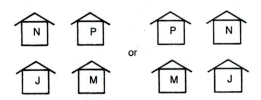

The third statement tells us that Sam and John live next door to one another. The sketches shown below represent the possibilities now that we have used all of the given facts.

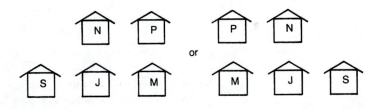

Now we can refer to the last two sketches to answer the questions.

1. John and Pat do not live on the same street.
2. Pat does not live next door to Sam.
3. Sam does not live behind Pat.
4. John lives next door to Marge. ◀

EXAMPLE 8   Jess, Gail, Richard, and Jane work at a university. One is a librarian, one is a vice president, one is a dean, and one is a teacher.

Jess and Richard ride to school with the dean.

Gail and Jane eat lunch with the teacher.

Jane works in the same building with the dean and the librarian.

Based on the preceding information, who is the dean?

Solution   In this problem, we can use a table to keep track of the given facts and reach a valid conclusion. We start by labeling four columns with the title of one of the positions (librarian, vice president, dean, and teacher). Under each heading we list the initials of the names of the four people (using Ja for Jane and Je for Jess). As we read through the given statements we will be able to eliminate some of the people from some of the positions.

LIBRARIAN	VICE PRESIDENT	DEAN	TEACHER
Je	Je	Je	Je
G	G	G	G
R	R	R	R
Ja	Ja	Ja	Ja

The first given statement tells us that Jess and Richard ride with the dean. Therefore we know that neither Jess nor Richard is the dean and we mark through their initials under the Dean column.

LIBRARIAN	VICE PRESIDENT	DEAN	TEACHER
Je	Je	J̶e̶	Je
G	G	G	G
R	R	R̶	R
Ja	Ja	Ja	Ja

From the second statement we know that Gail and Jane are not the teacher, so we cross out their initials in the Teacher column.

LIBRARIAN	VICE PRESIDENT	DEAN	TEACHER
Je	Je	J̶e̶	Je
G	G	G	G̶
R	R	R̶	R
Ja	Ja	Ja	J̶a̶

From the third given statement we know that Jane is not the dean or the librarian, so we cross out her initial in the Dean and Librarian columns.

LIBRARIAN	VICE PRESIDENT	DEAN	TEACHER
J̶e	J̶e	J̶e	J̶e
G	G	G	G̶
R	R	R̶	R
J̶a	Ja	J̶a	J̶a

The last table does not give the name of the person in every position, but it does have the answer to the question. In the Dean column there is only one person's name remaining. Therefore, Gail is the dean. ◀

EXAMPLE 9   At a university a poll was taken to determine how many students read *Newsweek* or the *Wall Street Journal*.

   200 students participated in the survey.

   96 students read *Newsweek*.

   94 students read the *Wall Street Journal*.

   80 students read both.

1. How many students read *Newsweek* but not the *Wall Street Journal*?
2. How many students read the *Wall Street Journal* but not *Newsweek*?
3. How many students read *Newsweek* or the *Wall Street Journal*?
4. How many students read neither *Newsweek* nor the *Wall Street Journal*?

Solution   Again we can represent the information with a drawing. This time the given information tells us how two groups of students (*Newsweek* readers and *Wall Street Journal* readers) are related to each other and to a larger group (the students participating in the poll) in which they are members. We draw one circle to represent the students who read *Newsweek* and another circle to represent the students who read the *Wall Street Journal*. Since some students read both, we draw the circles partially overlapping. We enclose the circles in a rectangle that represents all of the students participating in the poll. The region outside the circles but inside the rectangle represents those students who were polled but who don't read either *Newsweek* or the *Wall Street Journal*. The region outside the *Newsweek* circle but inside the *Wall Street Journal* circle represents those students polled who don't read *Newsweek* but do read the *Wall Street Journal*.

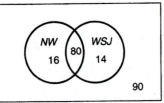

Now we enter the numbers given in the statement of the problem into the appropriate regions of the drawing. We start with the most inclusive information, which is in the last given statement. Eighty students read both periodicals, so we put 80 in the region where the two circles overlap. Since 94 students read the *wall Street Journal*, the circle for that periodical must have a total of 94 in it. However, the *Journal's* circle already has 80 in the part that overlaps *Newsweek's* circle. Thus there are only 94 − 80 = 14 *Journal* readers not accounted for, so we put 14 in the part of the *Journal's* circle that does not overlap *Newsweek's* circle. Similarly, we determine that there are 96 − 80 = 16 students in the part of *Newsweek's* circle that does not overlap the *Journal's* circle. If we total the number of students that we have now accounted for, 80 + 14 + 16, we find that we have accounted for only 110 of the 200 students polled. The remaining 200 − 110 = 90 students were polled but don't read either periodical. Therefore we enter 90 somewhere in the region inside the rectangle but outside the circles.

Drawings of the type we have just prepared are called Venn diagrams, which are discussed in the next section. Using this Venn diagram we can now answer the questions.

1. Sixteen students read *Newsweek* but not the *Wall Street Journal*.
2. Fourteen students read the *Wall Street Journal* but not *Newsweek*.

3. One hundred ten students read the *Wall Street Journal* or *Newsweek*. (Remember the meaning of "or"? The question means how many students read at least one of the two periodicals.)
4. Ninety students read neither periodical. ◀

When using deductive reasoning to solve problems, given statements often include quantifiers. A **quantifier** is a word or phase that tells "how many." The following table lists several quantifiers.

QUANTIFIER	MEANING	EXAMPLE
Some	At least one	Some people like me.
Most	More than 50%	Most people drink coffee.
A	At least one	A man came into my office.
All	The entire set	All men drive cars.
There exists	There is at least one	There exists a number whose absolute value is 3.
There is (are)	There is at least one	There are people who like math.
Many	More than one	Many people attended.

When a sentence has no quantifier, the quantifier usually is assumed to be "all." For example, we assume that the statement "Horses sleep standing up" means "All horses sleep standing up."

EXAMPLE 10   Assume that the following statements are true:

All people wearing shoes have brown hair.

Some people have red hair.

All people who have brown hair like ice cream.

People who have red hair like watermelon.

Sue has brown hair.

Based on the preceding five statements, determine whether each of the following conclusions is true or false.

1. Sue is wearing shoes.
2. Many people have red hair.
3. People who have brown hair like watermelon.
4. Sue likes ice cream.

Solution   1. The conclusion is false. Sue has brown hair, but there is no statement saying "All people with brown hair are wearing shoes."
2. Since "some" and "many" both mean "at least one," this conclusion is true.
3. This conclusion is not substantiated by the given statements.
4. This conclusion is true because Sue has brown hair and the third given statement tells us that all people with brown hair like ice cream. ◀

```
┌──────────────────────────────────────────────────┐
                     QUICK CHECK

   1.  In a hotel the positions of house detective, bell captain, and night
       manager are held by Tom, George, and Sam, but not respectively.
         a.  The house dectective is an only child and earns the least amount
             of money.
         b.  George married Sam's sister and earns more than the bell captain.
       Who is the night manager?

   2.  In a survey of 430 students it was found that

       270 students read one newspaper a day.

       220 students read one magazine a week.

       150 students read one newspaper a day and one magazine a week.

         a.  How many students in the survey do not read a newspaper or
             magazine?
         b.  How many students only read one newspaper a day and no
             magazines?

   3.  Assume that the following statements are true:

       All paint horses have spots.

       Some horses are black.

       All black horses like to eat alfalfa.

       Some horses like to eat oats.

       My horse has spots.

       Which of the following conclusions are valid?
         a.  My horse likes alfalfa.
         b.  My horse is a paint.
         c.  My horse likes oats.
         d.  None of the above is valid.
└──────────────────────────────────────────────────┘
```

2 INDUCTIVE REASONING

To successfully use inductive reasoning, you examine a situation, determine how the elements of the situation are related, and then predict what other elements will fit the same pattern.

EXAMPLE 11   Determine the next number in the sequence.

3, 5, 8, 13, 21, 34, 55, 89, . . .

Solution   This pattern of numbers is called a Fibonacci sequence. In a Fibonacci sequence the first two numbers are randomly chosen. Then any number after the second

```
┌──────────────────────────────────────────────────┐
 ANSWERS
 1. George   2. 90,120   3. (d)
└──────────────────────────────────────────────────┘
```

is the sum of the two preceding numbers. Thus the next number in this sequence is $55 + 89 = 144$. ◀

There are numerous patterns that can determine a sequence of numbers. For example, in the sequence

2, 6, 18, 54, . . .

each number after the first is obtained by multiplying the preceding number by 3. Or, in the sequence

2, 6, 10, 14, . . .

each number after the first is obtained by adding 4 to the preceding number. And finally, in the sequence

2, 4, 16, 256, . . .

each number is the square of the preceding term.
The possibilities are endless.

EXAMPLE 12   Determine the next number in the sequence.

5, 9, 17, 33, 62, 123, . . .

Solution   In this sequence each term is one less than twice the preceding term. Thus, the next number in the sequence is $2 \cdot 123 - 1 = 245$. ◀

In the following example we use inductive reasoning to find a pattern involving the arrangement of geometrical figures.

EXAMPLE 13   Determine the requirements for the missing figure.

Solution    As you examine the figures from left to right, notice that the inside shape in one figure becomes the outside shape in the next figure to the right. Or, if you examine the figures from right to left, the outside shape in one figure becomes the inside shape in the next figure to the left.

Thus if we examine the figures from left to right, we see that the missing figure must have a five-sided shape on the outside. If we examine the figures from right to left, we see that the missing shape must have a circle on the inside. Thus the figure shown in the margin is the missing figure. ◀

Another kind of geometrical pattern involves rotations of some aspect of the drawings. The next example presents a typical pattern of this kind.

EXAMPLE 14   Determine the requirements for the next figure in the following sequence.

Solution   There are several patterns at work here. First, there is a pattern involving the number of sides. The first figure has three sides, the second has four sides, and the third has five sides. Therefore, we expect the next figure to have six sides.

A second pattern is the number of dots in each figure. The first figure has one dot, the second figure has two dots, and the third figure has three dots. Therefore, we expect the next figure to have four dots.

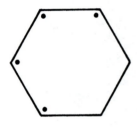

The third pattern is a little harder to figure out. Where do the dots go? We can see that they always go in adjacent corners of the figures, but where do they start? If you number the corners starting with the lower left corner in each drawing and proceed around the figure counterclockwise, you will see that the dot pattern starts in the first corner in the first figure, in the second corner in the second figure, and in the third corner in the third figure. Thus we expect the next figure to start its adjacent dots in the fourth corner from the lower left and continuing counterclockwise.

Based on the established patterns, the next figure would be like the one shown in the margin.   ◀

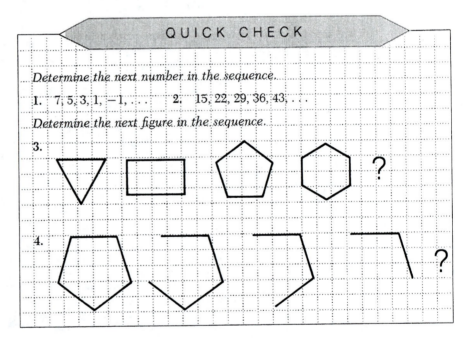

QUICK CHECK

Determine the next number in the sequence.

1.   7, 5, 3, 1, −1, . . .   2.   15, 22, 29, 36, 43, . . .

Determine the next figure in the sequence.

3.

4.

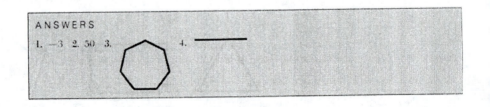

ANSWERS

1. −3 2. 50 3.   4. _____

ANSWERS

1. Assume that the following statements are true.

   If Raymond gets a raise, he will buy a new car or take a vacation.

   If Raymond sells one more major property, he will get a raise.

   The Lobo Ranch is a major property.

   If Raymond takes a vacation, he will go to London.

   If Raymond buys a new car, it will be a convertible.

   Raymond sells the Lobo Ranch and does not buy a convertible.

   Which of the following statements is true?
   a. Raymond does not get a raise.
   b. Raymond goes to London.
   c. Raymond buys a hardtop car.
   d. Raymond does not take a vacation.

   1.

   a. _____
   b. _____
   c. _____
   d. _____

2. A panel of four speakers is seated side by side at a table. One of the speakers is a freshman, one is a sophomore, one is a junior, and one is a senior. The senior sits at one end of the table. The sophomore sits immediately between the junior and the freshman. The freshman sits next to the senior. Which of the following is true?
   a. The junior sits at one end of the table.
   b. The freshman sits immediately between the junior and the sophomore.
   c. The sophomore sits next to the senior.
   d. The senior sits next to the junior.

   2.

   a. _____
   b. _____
   c. _____
   d. _____

3. A survey of the oral hygiene habits of 100 students revealed that 37 students floss daily, 62 students use a mouthwash daily, and 14 students both floss and use a mouthwash daily. Which of the following is *not* true?
   a. Sixteen students neither floss daily nor use a mouthwash daily.
   b. Forty-eight students use a mouthwash daily but do not floss daily.
   c. Twenty-three students floss daily but do not use a mouthwash daily.
   d. Eighty-five of the students floss or use a mouthwash daily.

   3.

   a. _____
   b. _____
   c. _____
   d. _____

4. What is the next number in the following sequence?
   $5, 2, -1, -4, \ldots$
   a. -7 b. 4 c. 2 d. -16

   4. _____

5. What is the next number in the following sequence?
   $2, 6, 18, 54, \ldots$
   a. 112 b. 58 c. 57 d. 162

   5. _____

6. What is the next number in the following sequence?
   $3, -6, 12, -24, \ldots$
   a. -33 b. -12 c. 48 d. 12

   6. _____

7. What is the next number in the following sequence?
   $2, 7, 3, 8, 4, \ldots$
   a. 5 b. 9 c. -1 d. 13

   7. _____

8. What is the next number in the following sequence?
   $2, 4, 7, 11, 16, \ldots$
   a. 22 b. 32 c. 18 d. 24

   8. _____

9. What number comes next in the following sequence?   9. _____
    1, 4, 10, 19, 31, . . .
    a. 33   b. 46   c. 42   d. 124

10. What number comes next in the following sequence?   10. _____
    1, 2, 2, 4, 8, . . .
    a. 9   b. 32   c. 12   d. 16

11. Refer to the figures shown and answer the question that follows.   11. _____

Which figure should come next?
a.   b.   c.   d.

12. Refer to the figures shown and answer the question that follows.   12. _____

        ?

Which figure should come next?
a.   b.   c.   d.

13. Refer to the figures shown and answer the question that follows.   13. _____

        ?

Which figure should come next?
a.   b.   c.   d.

14. Refer to the figures shown and answer the question that follows.

14. _____

     ?

Which figure should come next?

a.   b.   c.   d.

1.5 SETS AND SUBSETS

OBJECTIVES

▶ **1** Explain a well-defined set.

▶ **2** Symbolically represent sets and their elements using set notation.

▶ **3** Describe sets using the roster method and rule method.

▶ **4** Draw a Venn diagram.

▶ **5** Determine subsets and equal sets; define the Null set.

▶ **1** WELL-DEFINED SETS

Mathematics deals almost exclusively with various kinds of sets. A **set** is any well-defined collection. By "well-defined" we mean that you can determine without question whether or not a particular item belongs to a given set. For example, the set of all great books is not a well-defined set because the decision as to whether or not a particular book is a "great" book is a matter of opinion.

The set containing the numbers 1, 2, and 3 is well defined because, given any number, you can tell whether or not it belongs to the set. For example, if the given number is 1 or 2 or 3, it belongs to the set; any number other than 1, 2, or 3 does not belong to the set.

> ## QUICK CHECK
>
> *Determine whether each of the following is a well-defined set.*
>
> 1. Set A is the set of all letters in the English alphabet.
> 2. Set B is the set of all tall mountains.

▶ **2** REPRESENTING SETS AND THEIR ELEMENTS USING SET NOTATION

The items in a set are called **members** or **elements** of the set. When a set and its members are written using symbols, it is said to be written using **set notation**. Symbolically, a set is represented by enclosing its members within braces. Thus $\{1, 2, 3\}$ is read "the set containing 1, 2, and 3." The symbol $\in$ means that a given element "is an element of" or "is a member of" a particular set. Thus $3 \in \{1, 2, 3\}$ is read "3 is an element of the set containing 1, 2, and 3." When a slash is drawn through the symbol $\in$, the resulting symbol means that a given element "is not an element of" a particular set. Thus $6 \notin \{1, 2, 3\}$ is read "6 is not an element of the set containing the numbers 1, 2, and 3."

QUICK CHECK

Write each of the following sentences using set notation.

1. Nine is a positive integer.

2. Zero is not a positive integer.

The set of integers is $\{\ldots, -3, -2, -1, 0, 1, 2, 3, \ldots\}$.

 DESCRIBING A SET

One method for describing a set is called the **roster method** because it lists all the members of the set or uses a partial list to establish a pattern from which you can determine exactly what elements are in the set. The following sets are described using the roster method:

$A = \{1, 2, 3, 4, 5, 6, 7, 8, 9, 10\}$ — All of set A's elements are listed

$B = \{1, 2, 3, \ldots, 100\}$ — Set B's elements are partially listed, establishing a pattern; the three dots indicate that the pattern continues up to and including the last element, 100

$C = \{2, 4, 6, 8, \ldots\}$ — The dots indicate that the established pattern continues indefinitely

A second method used to describe sets is called the **rule method**, which describes the elements of the set by using a phrase or sentence. The following sets are described using the rule method:

$A = \{$names of the U.S. presidents$\}$

$B = \{$the students in your class whose grade point average is above 3.5$\}$

QUICK CHECK

Identify the method used to describe each set.

1. $A = \{\frac{1}{2}, \frac{1}{3}, \frac{1}{4}, \ldots, \frac{1}{10}\}$ 2. $B = \{$letters in the Greek alphabet$\}$

ANSWERS FOR QUICK CHECK AT TOP OF PAGE

1. $9 \in \{1, 2, 3, 4, \ldots\}$ 2. $0 \notin \{1, 2, 3, 4, \ldots\}$

ANSWERS FOR QUICK CHECK AT BOTTOM OF PAGE

1. Roster method   2. Rule method

VENN DIAGRAMS

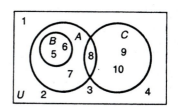

FIGURE 1.3

Sets are often displayed in a drawing that shows how they are interrelated. Such a drawing is called a **Venn diagram**. The Venn diagram in Figure 1.3 represents the following sets:

$$U = \{1, 2, 3, 4, 5, 6, 7, 8, 9, 10\}$$
$$A = \{5, 6, 7, 8\}$$
$$B = \{5, 6\}$$
$$C = \{8, 9, 10\}$$

Set U, represented by the rectangular region, is called the **universal set** and contains all of the elements and all of the sets under discussion.

QUICK CHECK

Use the roster method to describe the four sets in the following Venn diagram.

SUBSETS, EQUAL SETS, AND THE NULL SET

Set A is a **subset** of set B if and only if all of the elements of A are also elements of B.

The symbol $\subseteq$ means "is a subset of" and the symbol $\nsubseteq$ means "is not a subset of." Thus $A \subseteq B$ is read "A is a subset of B" and $A \nsubseteq B$ is read "A is not a subset of B." Consider the sets

$$A = \{1, 2, 3, 4, 5\}$$

and

$$B = \{0, 1, 2, 3, 4, 5, 6, 7, 8\}$$

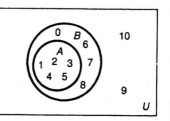

FIGURE 1.4

that are illustrated in the Venn diagram in Figure 1.4. Note that all of the elements in set A are also elements of set B. Thus we can conclude that A is a subset of B and write $A \subseteq B$.

ANSWERS

1. $A = \{1, 2, 4, 5, 6, 7, 8, 12\}$ 2. $B = \{3, 4, 5, 6, 7, 9, 10\}$ 3. $C = \{3, 4, 5, 12, 14\}$
4. $U = \{1, 2, 3, 4, 5, 6, 7, 8, 9, 10, 11, 12, 13, 14\}$

Every set is a *subset of itself*. For example, set A is a subset of itself (written $A \subseteq A$) because all of the elements of A are elements of A. Sometimes we say that such a set is an **improper subset** of itself.

Set A is called a **proper subset** of set B iff $A \subseteq B$ but $B \nsubseteq A$. That is, all of the elements in A must be elements of B, and B must have at least one element that A does not have. The symbol $\subset$ means "is a proper subset of." Thus $A \subset B$ is read "A is a proper subset of B."

In Figure 1.5, A is a proper subset of B ($A \subset B$) because all of A's elements are also elements of B, and B has at least one element that A does not have.

Two sets that have exactly the same elements are called **equal sets**. Therefore, sets K and L are equal iff $K \subseteq L$ and $L \subseteq K$. Note that when a particular set is described by listing all of its elements, the order in which the elements are listed is immaterial. For example, $\{1, 2, 3\} = \{3, 1, 2\}$. This is a direct result of the definition of equal sets as any two sets that have the same elements.

The **null set** is a set having no elements and is often called the **empty set**. The symbols used to describe this set are a pair of braces with no elements between them, $\{ \ \}$, or the Greek letter phi, ϕ. By agreement, the null set is considered a proper subset of every set except itself, and it is an improper subset of itself. That is, for any nonempty set A, $\phi \subset A$, whereas $\phi \subseteq \phi$.

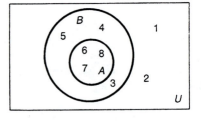

FIGURE 1.5

QUICK CHECK

Consider the following sets:

$A = \{$positive integers$\}$

$B = \{$integers$\}$

$C = \{$negative integers$\}$

$D = \{$integers greater than one and less than 2$\}$

$E = \{1, 2, 3, 4, \ldots\}$

1. Is $A \subseteq B$?   2. Is $A \subset B$?   3. Is $B \subseteq C$?   4. Is $A = E$?
5. Which set is empty?

EXERCISE 1.5

A

ANSWERS

1. What does it mean to say that a set is well defined?

1. _____

2. Define the null set.

2. _____

3. If $A = \{1, 2, 3\}$, name eight subsets (proper and improper) of A.

3. _____

4. If $B = \{3, 4, 6\}$, name eight subsets (proper and improper) of B.

4. _____

Give two examples that illustrate each of the following.

5. Subsets

5. _____

6. Three dots in a listing

6. _____

7. Equal sets

7. _____

8. The rule method for describing sets

8. _____

9. The roster method for describing sets

9. _____

10. Proper subsets

10. _____

Identify the method used to describe each of the following sets.

11. $\{1, 2, 3, 4 \ldots\}$

11. _____

12. $\{3, 5, 7\}$

12. _____

13. $\{1, 5, 10\}$

13. _____

14. $\{1, 3, 5, 7, \ldots, 11\}$

14. _____

15. {people under 20 years old}

15. _____

16. {dogs that walk on two legs}

16. _____

Write each of the following using mathematical symbols.

17. Describe the set of all possible scores in football using the roster method.

17. _____

18. Describe the set of all letter grades used in your school using the roster method.

18. _____

19. Describe the set {dogs, cats} using the rule method.

19. _____

20. Describe the set $\{0, 1, 2, 3, 4, 5, 6, 7, 8, 9, 10\}$ using the rule method.

20. _____

Consider the Venn diagram shown and answer each of the following true or false.

21. $B = \{4, 5\}$

21. _____

22. $B = \{4, 5, 6, 7\}$

22. _____

23. $A = \{6, 7\}$

23. _____

24. $A = \{4, 5, 6, 7\}$

24. _____

25. $U = \{1, 2, 3\}$

25. _____

26. $U = \{4, 5, 6, 7\}$

26. _____

27. $U = \{1, 2, 3, 4, 5, 6, 7\}$

27. _____

28. $4 \in A$

28. _____

29. $4 \in U$ 29. _____

30. $4 \notin A$ 30. _____

31. $4 \notin U$ 31. _____

32. $5 \notin A$ 32. _____

33. $5 \notin B$ 33. _____

34. $A \subseteq B$ 34. _____

35. $A \subset B$ 35. _____

36. $B \subset A$ 36. _____

37. $B \subseteq A$ 37. _____

38. $6 \in A$ 38. _____

39. $6 \in B$ 39. _____

40. $6 \in U$ 40. _____

41. $B \in U$ 41. _____

42. $B \subseteq U$ 42. _____

43. $A \subseteq U$ 43. _____

44. $A \subset U$ 44. _____

45. $4 \in B$ 45. _____

46. $\phi \subset A$ 46. _____

47. $\phi \subset B$ 47. _____

48. $\phi \subset U$ 48. _____

B

Answer each of the following true or false.

49. $\phi \subset \{0\}$ 49. _____

50. $\phi \subseteq \{0\}$ 50. _____

51. $\phi = \{0\}$ 51. _____

52. $\phi \subset \phi$ 52. _____

53. $\phi \subseteq \phi$ 53. _____

1.6 INTERSECTION OF SETS, DISJOINT SETS, AND UNION OF SETS

OBJECTIVES

1 ▶ Find the intersection of sets.

2 ▶ Define and identify disjoint sets.

3 ▶ Find the union of sets.

▶ INTERSECTION OF SETS

Two sets intersect if and only if they have at least one element in common. If two sets have no elements in common, their intersection is empty and we use the null set to represent the intersection (or lack of intersection).

> **INTERSECTION OF SETS**
>
> The **intersection** of two sets is a set containing the elements the two sets have in common.

The following sets A, B, and U are shown in Figure 1.6.

$$U = \{0, 1, 2, 3, \ldots, 20\}$$
$$A = \{1, 2, 3, 4, 5, 6, 7, 8, 9, 10\}$$
$$B = \{8, 9, 10, 11, 12, 13, 14, 15\}$$

The intersection of A and B is $\{8, 9, 10\}$. Note that the intersection of sets A and B is a set, not just a listing of shared elements.

FIGURE 1.6

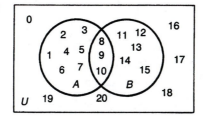

The symbol $\cap$ is used to represent intersection. For the sets in Figure 1.6 we write $A \cap B = \{8, 9, 10\}$, which is read "A intersect B is the set containing 8, 9, and 10" or "the intersection of sets A and B is the set containing 8, 9, and 10." Just as addition was an operation performed on two numbers, finding the intersection is an operation performed on two sets and is called a **set operation**, and there are others.

For a number to be included in the intersection of two sets, it must appear in both of the given sets. If we say "Find the set of numbers x such that x belongs to $\{1, 2, 3\}$ and x is even," you are being asked to find the intersection of the following sets.

$$\underbrace{\{1, 2, 3\}}_{x \text{ is in this set}} \quad \underset{\underset{\text{and}}{\downarrow}}{\cap} \quad \underbrace{\{\ldots, -4, -2, 0, 2, 4, 6, \ldots\}}_{x \text{ is an even number}} = \{2\}$$

▶2 DISJOINT SETS

If two sets do not intersect (i.e., their intersection is the null set), then they are
called **disjoint sets**.

DISJOINT SETS
_____

Two sets are **disjoint** if and only if they have no elements in common.

In Figure 1.7 sets A and B are disjoint sets; therefore $A \cap B = \phi$.

FIGURE 1.7

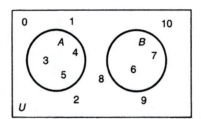

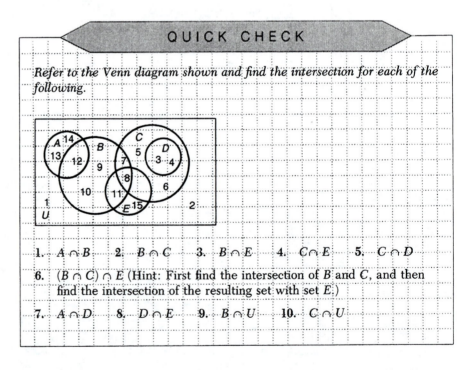

QUICK CHECK

*Refer to the Venn diagram shown and find the intersection for each of the
following.*

1. $A \cap B$ 2. $B \cap C$ 3. $B \cap E$ 4. $C \cap E$ 5. $C \cap D$

6. $(B \cap C) \cap E$ (Hint: First find the intersection of B and C, and then
find the intersection of the resulting set with set E.)

7. $A \cap D$ 8. $D \cap E$ 9. $B \cap U$ 10. $C \cap U$

▶3 THE UNION OF SETS

Another operation that can be performed with sets is called the union of sets.

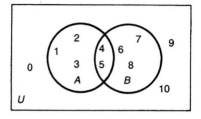

FIGURE 1.8

UNION OF SETS

If all of the elements of two sets are combined to form a new set (without duplicating elements), then the resulting set is the **union** of the two original sets.

Figure 1.8 shows the following sets:

$$U = \{0, 1, 2, 3, 4, \ldots, 10\}$$
$$A = \{1, 2, 3, 4, 5\}$$
$$B = \{4, 5, 6, 7, 8\}$$

The symbol $\cup$ is used to represent union. The phrase $A \cup B$ is read "A union B" or "the union of A and B." To find the union of sets A and B, we form a new set containing all of A's elements and all of B's elements.

$$A \cup B = \{1, 2, 3, 4, 5\} \cup \{4, 5, 6, 7, 8\}$$
$$= \{1, 2, 3, 4, 5, 6, 7, 8\}$$

Notice that the elements are not duplicated in the union of two sets, even though they appear in each set. Finding the union of two sets is another set operation.

The word "or" describes the union of two sets. If we say "Find the set of numbers x such that x is a positive even integer less than 10 or x is a positive odd integer less than 15," we must find the union of the following sets.

$$\underbrace{\{2, 4, 6, 8\}}_{\substack{x \text{ is a positive} \\ \text{even number less} \\ \text{than 10}}} \quad \underset{\substack{\downarrow \\ \text{or}}}{\cup} \quad \underbrace{\{1, 3, 5, 7, 9, 11, 13\}}_{\substack{x \text{ is a positive} \\ \text{odd number less} \\ \text{than 15}}} = \{1, 2, 3, 4, 5, 6, 7, 8, 9, 11, 13\}$$

QUICK CHECK

Refer to the Venn diagram shown and find the union for each of the following.

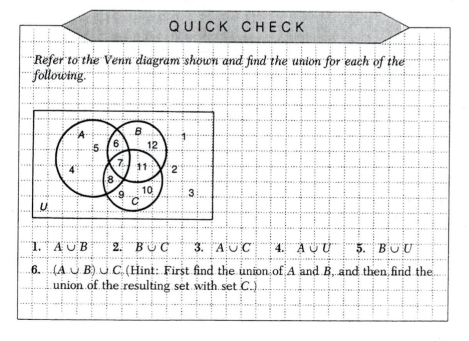

1. $A \cup B$ 2. $B \cup C$ 3. $A \cup C$ 4. $A \cup U$ 5. $B \cup U$

6. $(A \cup B) \cup C$ (Hint: First find the union of A and B, and then find the union of the resulting set with set C.)

ANSWERS

1. $\{4, 5, 6, 7, 8, 11, 12\}$ 2. $\{6, 7, 8, 9, 10, 11, 12\}$ 3. $\{4, 5, 6, 7, 8, 9, 10, 11\}$ 4. U 5. U
6. $\{4, 5, 6, 7, 8, 9, 10, 11, 12\}$

EXAMPLE 15

Refer to the Venn diagram shown and find the intersection or union for each of the following.

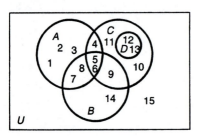

1. $A \cup B$ 2. $A \cap C$ 3. $D \cap C$ 4. $D \cup C$

For Problems 5–8, first find the intersection or union of the sets within the parentheses; then find the intersection or union of the resulting set with the third set.

5. $(A \cap B) \cap C$ 6. $(A \cup B) \cup C$ 7. $(A \cap B) \cup C$ 8. $(A \cup B) \cap C$

Solution

1. The union of sets A and B is the set containing all of their elements. Thus

$$A \cup B = \{1, 2, 3, 4, 5, 6, 7, 8, 9, 14\}$$

2. The intersection of sets A and C is a set containing the elements they have in common. Thus

$$A \cap C = \{4, 5, 6\}$$

3. The intersection of D and C is $\{12, 13\}$, or set D, because all of D's elements are in C.

4. The union of D and C is $\{4, 5, 6, 9, 10, 11, 12, 13\}$, or set C.

5. Find the intersection of A and B first: $A \cap B = \{5, 6, 7, 8\}$. Then find the intersection of this set with set C, which is $\{5, 6\}$.

6. Find A union B first $A \cup B = \{1, 2, 3, 4, 5, 6, 7, 8, 9, 14\}$. The union of this set with set C is

$$\{1, 2, 3, 4, 5, 6, 7, 8, 9, 10, 11, 12, 13, 14\}$$

7. $A \cap B = \{5, 6, 7, 8\}$. The union of this set with set C is

$$\{4, 5, 6, 7, 8, 9, 10, 11, 12, 13\}$$

8. The union of A and B is $\{1, 2, 3, 4, 5, 6, 7, 8, 9, 14\}$. The intersection of this set with set C is $\{4, 5, 6, 9\}$.   ◀

◄ E X E R C I S E 1 . 6

A

For Exercises 1–4 give two examples for each.

1. Intersection of two sets

2. Disjoint sets

3. Union of two sets

4. Intersection of two sets having a single element in common

5. If A is a proper subset of B, then $A \cap B$ must be equal to which set, A or B?

6. If A is a proper subset of B, then $A \cup B$ must be equal to which set, A or B?

1. _____

2. _____

3. _____

4. _____

5. _____

6. _____

For Exercises 7–19 refer to the accompanying Venn diagram.

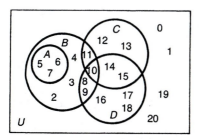

7. Are sets A and C disjoint?

8. Are sets A and D disjoint?

9. Are sets B and C disjoint?

10. Are sets C and D disjoint?

11. Find $A \cap B$.

12. Find $B \cap C$.

13. Find $(A \cap B) \cap C$.

14. Find $A \cup B$.

15. Find $B \cup C$.

16. Find $U \cap A$.

17. Find $U \cup A$.

18. Find $(B \cap C) \cup D$.

19. Find $(C \cap D) \cup B$.

7. _____

8. _____

9. _____

10. _____

11. _____

12. _____

13. _____

14. _____

15. _____

16. _____

17. _____

18. _____

19. _____

B

There are 1400 students in our school. Of these students 50 are presently taking English, history, and mathematics; 200 are taking English and history only; 150 are taking history and math only; and 350 are taking English and math only. There are totals of 650 taking English; 500 taking history; and 600 taking math.

20. Draw a Venn diagram representing the number of people in the school taking these three subjects.

21. How many students are taking math only?

22. How many students are taking English only?

23. How many students are taking history only?

24. How many students are not taking English, history, or math?

20. _____

21. _____

22. _____

23. _____

24. _____

A school has three organizations to which a total of 36 seniors could belong. There are only 2 seniors that belong to all three clubs. Exactly 3 seniors belong to both clubs A and B; exactly 8 seniors belong to both clubs B and C; and exactly 5 seniors belong to both clubs A and C. Club A has a total of 13 seniors; club B has a total of 13 seniors; and club C has a total of 16 seniors.

25. Draw a Venn diagram representing the membership in each club.

26. How many seniors belong to club A only?

27. How many seniors belong to club B only?

28. How many seniors belong to club C only?

29. How many seniors belong to clubs A or B only?

30. How many seniors do not belong to a club?

25. _____

26. _____

27. _____

28. _____

29. _____

30. _____

Explain why each of the following is not a mathematical definition.

ANSWERS

1. A wren is a bird.

1. _____

2. If it is a square, then it is a rectangle.

2. _____

3. A circle is a round object.

3. _____

Answer each of the following true or false.

4. Circular definitions are used in mathematics.

4. _____

5. An undefined term has no real meaning.

5. _____

6. A "set" is an undefined term.

6. _____

7. A mathematical definition must place the word or phrase being defined into a set and then show how it can be distinguished from all other members of the set.

7. _____

8. "Element" and "member" are synonyms.

8. _____

9. Disjoint sets can have elements in common.

9. _____

10. If a statement is true, then its converse is also true.

10. _____

11. For a biconditional statement to be true the corresponding conditional and converse statements must be true.

11. _____

12. The "if clause" of a conditional statement is the hypothesis.

12. _____

13. A biconditional statement may be written as two conditional statements combined with the word "or."

13. _____

14. Mathematical definitions may be written as conditional statements.

14. _____

15. Properties or axioms are rules that are accepted without justification.

15. _____

16. Rules or theorems are statements that can be justified by properties, definitions, or previously proven rules.

16. _____

17. A slash through an operation symbol negates it.

17. _____

18. A slash through a relationship symbol negates it.

18. _____

19. When using the roster method to describe a set, you must list all of the set's elements.

19. _____

20. When using the rule method to describe a set, you must show a list of all of the elements of the set.

20. _____

21. If $A \subset B$, then $A \cap B = A$.

21. _____

22. If $A \subseteq B$, then $A \cap B = B$.

22. _____

23. If $A \subset B$, then $A \cup B = A$.

23. _____

24. If $A \subseteq B$, then $A \cup B = B$.

24. _____

25. If two sets intersect, then they must have two or more elements in common.

25. _____

26. If $A = \{1, 2, 3\}$ and $B = \{2, 3, 4\}$, then $A \cup B = \{1, 2, 3, 4\}$.

26. _____

27. If $A \subset B$, then A and B cannot be equal sets.

27. _____

Write each word, phrase, or sentence using mathematical symbols.

28. Is an element of   28. _____

29. Is not an element of   29. _____

30. The set containing 8   30. _____

31. Set A is a proper subset of set B.   31. _____

32. Z is not a subset of R.   32. _____

33. Union   33. _____

34. Intersection   34. _____

35. The intersection of sets P and Q 35. _____

Define or explain each of the following.

36. The roster method for describing sets   36. _____

37. The rule method for describing sets   37. _____

38. Equal sets   38. _____

39. Disjoint sets   39. _____

Refer to the following sets and answer Exercises 40–47 true or false. If your answer is false, explain why.

$$A = \{0, 1, 2, 3, \ldots, 9\} \qquad B = \{0, 1, 2, 3, \ldots, 10\}$$
$$C = \{1, 2, 3, 4, \ldots\} \qquad D = \{2, 4, 6, 8, \ldots\}$$
$$E = \{0, 1, 3, 5, 7, \ldots\} \qquad F = \{1, 2, 3\}$$

40. $A \subseteq C$ 40. _____

41. A and B are disjoint.   41. _____

42. $A \cup B = F$ 42. _____

43. $A \cap B = \{0\}$ 43. _____

44. D and E do not intersect.   44. _____

45. $D \subsetneq C$ 45. _____

46. $F \subseteq D$ 46. _____

47. $A \cap C = A$ 47. _____

48. Draw a Venn diagram to represent the following sets: 48. _____

$$U = \{1, 2, 3, \ldots, 10\}$$
$$A = \{2, 3, 5, 6\}$$
$$B = \{1, 2, 3, 4\}$$
$$C = \{1, 3, 5, 7, 8\}$$

49. Find the set whose elements are represented by x if   49. _____
$x \in \{1, 2, 3, 4\}$ or $x \in \{3, 4, 5\}$.

Exercises 50–60 refer to the following sets:

$A = \{1, 2, 3, 4, \ldots, 10\}$ $\qquad B = \{1, 3, 5, 7, 9\}$

$C = \{2, 4, 6, 8, 10\}$ $\qquad D = \{4, 8\}$

$E = \{0, 1, 2, 3, \ldots, 9\}$ $\qquad F = \{1, 2, 3, 4, \ldots\}$

$G = \{0\}$

50. Which two sets have 0 as an element? $\qquad$ 50. _____

51. Is $A = E$? $\qquad$ 51. _____

52. The union of sets B and C is equal to which set? $\qquad$ 52. _____

53. Which set has A as a proper subset? $\qquad$ 53. _____

54. $A \cap C$ is equal to which set? $\qquad$ 54. _____

55. Is $D \subseteq E$? $\qquad$ 55. _____

56. Is $1 \in G$? $\qquad$ 56. _____

57. Name two disjoint sets. $\qquad$ 57. _____

58. G is a proper subset of which set? $\qquad$ 58. _____

59. Is $E \subset F$? $\qquad$ 59. _____

60. Is $8 \in A$? $\qquad$ 60. _____

61. There are 34 students in a school who play sports after school. Some play basketball, some play baseball, and some play volleyball. There are 3 students who play all three sports; 5 students play only volleyball; 1 student plays only baseball and volleyball; and 2 students play only basketball and baseball. A total of 20 students play baseball; a total of 14 students play basketball; and a total of 16 students play volleyball. How many just play basketball? $\qquad$ 61. _____

62. Find the next number in the sequence. $\qquad$ 62. _____

8, 11, 14, 17, 20, ...

63. Determine the missing figure. $\qquad$ 63. _____

ANSWERS:

1. Explain what is meant by "undefined terms" in mathematics.

1. _____

2. What are the requirements for a mathematical definition?

2. _____

3. What are properties and rules and how do they differ?

3. _____

4. In what form should a mathematical definition be written?

4. _____

5. Explain the difference between conditional and biconditional statements.

5. _____

Write the following two statements as conditional statements.

6. A dog is a domestic animal.

6. _____

7. A number divisible by 2 is even.

7. _____

Write the following two statements as biconditional statements.

8. A number divisible by 3 is a multiple of 3.

8. _____

9. The numbers 0, 1, 2, 3, 4, 5, 6, 7, 8, or 9 are called digits.

9. _____

10. If a biconditional statement is true, what do you know about its statement and converse?

10. _____

11. Consider the statement "Weslie will buy a white car or a four-door car." If the statement is true, what can you conclude?

11. _____

12. Consider the statement "Carla is going to San Antonio and to Houston." If the statement is true, what can you conclude?

12. _____

13. Which symbol represents the phrase "is an element of"?
   a. $\subset$ b. $\subseteq$ c. $\in$ d. $\notin$ e. None of these

13. _____

14. Which symbol is used to represent the intersection of two sets?
   a. $\cup$ b. $\cap$ c. $\subset$ d. $\subseteq$ e. None of these

14. _____

15. If $A = \{1, 2\}$ and $B = \{3, 4\}$, which word or phrase best describes the relationship between sets A and B?
   a. Subsets   b. Proper subsets   c. Disjoint sets
   d. Intersecting sets   e. None of these

15. _____

16. Which of the following represents the intersection of sets P and Q?
   a. $P \cap Q$ b. $P \cup Q$ c. $P \subseteq Q$ d. $P \subset Q$
   e. None of these

16. _____

17. Which symbol represents union?
   a. $\cup$ b. $\cap$ c. $\subset$ d. $\subseteq$ e. None of these

17. _____

18. Draw a Venn diagram representing the following sets:

   $U = \{1, 2, 3, \ldots, 15\}$

   $A = \{1, 2, 3, 4, 5, 6, 7, 8\}$

   $B = \{1, 2\}$

   $C = \{5, 6, 7, 9, 10, 12, 13\}$

   $D = \{6, 7, 8, 9, 10, 11\}$

18. _____

For Problems 19–26 refer to the sets in Problem 18 and find the specified sets.

19. $A \cap B$

19. _____

20. $A \cap C$

20. _____

21. $A \cap D$ 21. _____

22. $D \cap U$ 22. _____

23. $B \cap U$ 23. _____

24. $A \cup B$ 24. _____

25. $A \cup C$ 25. _____

26. $A \cup U$ 26. _____

Problems 27–30 refer to the sets in Problem 18.

27. Is $A \subset B$?   27. _____

28. Is $B \subset A$?   28. _____

29. Is $A \subseteq U$?   29. _____

30. Is $\phi \subset B$?   30. _____

31. In a poll of 100 first-semester college students the following information   31. _____
    was determined.

    44 took English

    36 took history

    31 took math

    10 took English and math

    12 took history and math

    7 took English and history

    3 took English, math, and history

    How many of the 100 students did *not* take English, history, or math in
    thier first semester?

32. Find the next number in the series.   32. _____

    7, 9, 16, 25, 41, . . .

33. Find the next figure in the pattern.   33. _____

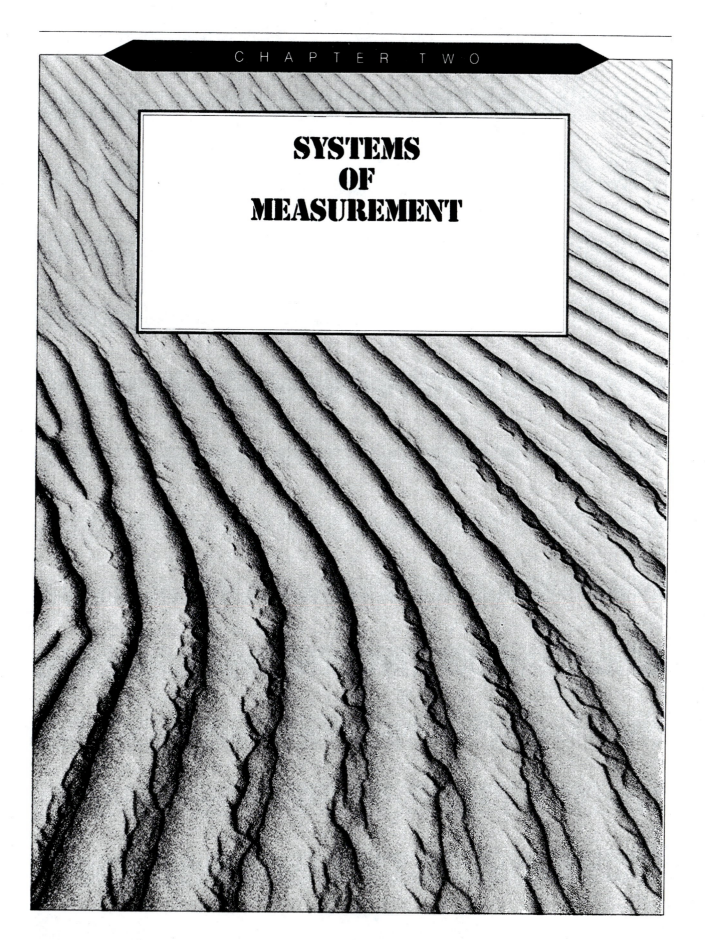

SYSTEMS
OF
MEASUREMENT

2.1 METRIC AND AMERICAN UNITS OF LENGTH

OBJECTIVES

▶ 1 Convert units of length in the American system.

▶ 2 Convert units of length in the metric system.

▶ 3 Convert units of length between the American and metric systems.

The concept of measurement is probably as old as the concept of number. In fact, many believe that numbers were invented to communicate the concept of measurement. The primitive measurements of antiquity have given way to the precise twentieth century science of measurement called metrology. In this book, we will assume our measurements are exact. Measurements in the industrial world are not exact and offer a real challenge. For example, consider the following problem.

Suppose you needed to find the width of your desk correct to five decimal places. Where would you begin? What instrument would you use and how could you guarantee that your instrument was that accurate? These are only a few of the questions that a metrologist (measurement scientist) must answer daily.

In this chapter we study two systems of measurement—the **American system** (or **English system**) and the **metric system**. The United States is the only major country in the world that uses the American system. All other major countries use the metric system.

▶ AMERICAN SYSTEM UNITS OF LENGTH

In the American system the basic units of measure for length are the inch, foot, yard, and mile. Equivalent units and abbreviations for these fundamental units of length are as follows.

AMERICAN SYSTEM UNITS OF LENGTH	
Equivalent Units	Abbreviations
12 inches = 1 foot	inch = in.
36 inches or 3 feet = 1 yard	foot = ft
5280 feet = 1 mile	yard = yd
	mile = mi

Feet and inches may also be abbreviated by placing a ′ or a ″ mark, respectively, after the number. Thus 2′ is read "two feet" and 2″ is read "two inches."

The process of changing one unit of measure to another unit of measure is called **unit conversion**. We use two mathematical concepts when we convert from one unit to another. Recall that any nonzero number divided by itself is one. If we know that $a = b$ and a and b are not zero, then we can conclude that

$$\frac{a}{b} = 1 \quad \text{or} \quad \frac{b}{a} = 1$$

Therefore, since 12 inches = 1 foot,

$$\frac{12\text{ inches}}{1\text{ foot}} = 1 \qquad \text{or} \qquad \frac{1\text{ foot}}{12\text{ inches}} = 1$$

From this concept it follows that since 1 inch = 1 inch,

$$\frac{1\text{ inch}}{1\text{ inch}} = 1$$

Now let's see how we use these concepts to convert from one unit to another.

EXAMPLE 1   Convert 15 inches to yards.

Solution   To convert 15 inches to yards, we multiply the given units (15 inches) by the form of the number one necessary to end up with yards. Since 1 yard = 36 inches, we multiply 15 inches by

$$\frac{1\text{ yard}}{36\text{ inches}} = 1$$

Given units = Given units · Conversion factor

$$15\text{ inches} = \frac{15}{1} \cdot \frac{\text{inches}}{1} \cdot \frac{1\text{ yard}}{36\text{ inches}} \qquad \text{Putting given units (inches) over 1}$$

$$= \frac{15}{36} \cdot \frac{\text{yards}}{1} \cdot \frac{1\text{ inch}}{1\text{ inch}} \qquad \text{Rearranging products}$$

$$= \frac{15}{36} \cdot \frac{\text{yards}}{1} \cdot 1 \qquad \text{Multiplying; reducing fraction}$$

$$= \frac{5}{12}\text{ yards}$$

Thus, 15 inches = $\frac{5}{12}$ yards.   ◀

Each time you use a conversion factor, place the numbers corresponding to the conversion facts with the appropriate units. For instance, in Example 1 we used the conversion fact 36 inches = 1 yard, so we placed the 36 with the inches and the 1 with the yard in the conversion factor

$$\frac{1\text{ yard}}{36\text{ inches}}$$

You may wonder why we used

$$\frac{1\text{ yard}}{36\text{ inches}} \qquad \text{instead of} \qquad \frac{36\text{ inches}}{1\text{ yard}}$$

We were looking ahead and could see that we would be able to rearrange the products to create a factor of 1 from

$$\frac{1\text{ inch}}{1\text{ inch}}$$

and thus have only yards remaining on the right.

Now let's convert 15 yards to inches. Notice the difference in the arrangement of the conversion factor.

EXAMPLE 2   Convert 15 yards to inches.

Solution   Since 36 inches = 1 yard,

$$\frac{36 \text{ inches}}{1 \text{ yard}} = 1$$

We multiply by this form for 1, rearrange the products, and have only inches remaining on the right.

$$15 \text{ yards} = \frac{15}{1} \cdot \frac{\text{yards}}{1} \cdot \frac{36 \text{ inches}}{1 \text{ yard}} \qquad \text{Put given units (yards) over 1}$$

$$= \frac{15}{1} \cdot \frac{36 \text{ inches}}{1} \cdot \frac{1 \text{ yard}}{1 \text{ yard}} \qquad \text{Rearranging products}$$

$$= \frac{15}{1} \cdot \frac{36}{1} \cdot \frac{\text{inches}}{1} \cdot 1$$

$$= 540 \text{ inches}$$

Thus 15 yards = 540 inches.   ◀

At times you may have to multiply the given units by more than one factor of the number one to make the desired conversion.

EXAMPLE 3   Change 23,000,000 inches to miles.

Solution   To change 23,000,000 inches to miles, we first convert from inches to feet and then from feet to miles.

$$23{,}000{,}000 \text{ inches} = 23{,}000{,}000 \, \frac{\text{inches}}{1} \cdot \frac{1 \text{ foot}}{12 \text{ inches}}$$

$$= \frac{23{,}000{,}000}{12} \text{ feet} \qquad \begin{array}{l} \text{Rearranging products;} \\ \text{dividing out } \dfrac{\text{inches}}{\text{inches}} \end{array}$$

$$= \frac{23{,}000{,}000}{12} \cdot \frac{\text{feet}}{1} \cdot \frac{1 \text{ mile}}{5280 \text{ feet}}$$

$$= \frac{23{,}000{,}000}{12 \cdot 5280} \text{ miles} \qquad \begin{array}{l} \text{Rearranging products;} \\ \text{dividing out } \dfrac{\text{feet}}{\text{feet}} \end{array}$$

$$= 363 \text{ miles} \qquad \begin{array}{l} \text{Rounding to nearest} \\ \text{mile} \end{array}$$

It is possible to work the problem in Example 3 without using so many steps. Consider the following conversion.

$$23{,}000{,}000 \text{ inches} = 23{,}000{,}000 \, \frac{\text{inches}}{1} \cdot \frac{1 \text{ foot}}{12 \text{ inches}} \cdot \frac{1 \text{ mile}}{5280 \text{ feet}} \qquad \begin{array}{l} \text{Using slash} \\ \text{marks to show} \\ \text{which units} \\ \text{divided out,} \\ \text{creating factors} \\ \text{of 1} \end{array}$$

$$= \frac{23{,}000{,}000}{12 \cdot 5280} \qquad \begin{array}{l} \text{Multiplying} \\ \text{fractions} \end{array}$$

$$= 363 \text{ miles} \qquad \begin{array}{l} \text{Rounding to} \\ \text{nearest mile} \end{array} \quad ◀$$

QUICK CHECK

Perform the indicated unit conversions. Where applicable round answers to the hundredths place.

1. 20 yds = _____ in.   2. 36 ft = _____ yd

3. 236,175 ft = _____ mi   4. 18 mi = _____ ft

▶ 2 METRIC SYSTEM UNITS OF LENGTH

Units of measure in the metric system are based on factors of 10. The most commonly used units of measure for length are the meter, millimeter, centimeter, and kilometer, although the metric system has three other units of measure.

METRIC SYSTEM UNITS OF LENGTH

Equivalent Units	Abbreviations
10 hectometers = 1 kilometer	kilometer = km
10 dekameters = 1 hectometer	hectometer = hm
10 meters = 1 dekameter	dekameter = dam
10 decimeters = 1 meter	meter = m
10 centimeters = 1 decimeter	decimeter = dm
10 millimeters = 1 centimeter	centimeter = cm
	millimeter = mm

In the metric system, each category of metric measure (length, mass, capacity, etc.) has a fundamental unit. For length the fundamental unit is the meter, for mass it is the gram, and for capacity it is the liter. Each category then uses the same prefixes, which have the following meanings.

METRIC SYSTEM PREFIXES

Prefix	Meaning
kilo	1000
hecto	100
deka	10
deci	$\frac{1}{10}$
centi	$\frac{1}{100}$
milli	$\frac{1}{1000}$

Thus for the meter,

$$1 \text{ kilometer} = 1000 \text{ meters} \qquad 1 \text{ decimeter} = \tfrac{1}{10} \text{ meter}$$
$$1 \text{ hectometer} = 100 \text{ meters} \qquad 1 \text{ centimeter} = \tfrac{1}{100} \text{ meter}$$
$$1 \text{ dekameter} = 10 \text{ meters} \qquad 1 \text{ millimeter} = \tfrac{1}{1000} \text{ meter}$$

ANSWERS

1. 720   2. 12   3. 44.73   4. 95,040

Since each consecutive metric unit of measure is related by a factor of 10, it is possible to construct a metric ladder and count up the ladder by factors of 10 (10, 100, 1000, 10,000, etc.) to convert from one unit to another.

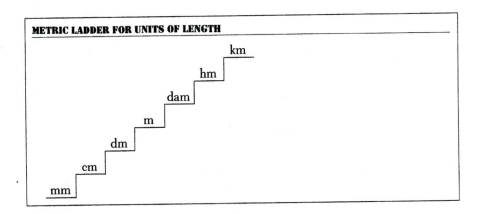

METRIC LADDER FOR UNITS OF LENGTH

The metric ladder can be used to determine relationships between metric units. Suppose we want to know the relationship between centimeters and hectometers. Start at the smallest of the two units, centimeters, and count up the metric ladder to hectometers by factors of 10.

$$
\begin{array}{ll}
10,000 \rightarrow \text{hm} & \text{since } 10,000 \text{ cm} = 1 \text{ km} \\
1000 \rightarrow \text{dam} & \text{since } 1000 \text{ cm} = 1 \text{ dam} \\
100 \rightarrow \text{m} & \text{since } 100 \text{ cm} = 1 \text{ m} \\
10 \rightarrow \text{dm} & \text{since } 10 \text{ cm} = 1 \text{ dm} \\
\text{Start} \rightarrow \text{cm} &
\end{array}
$$

Thus 10,000 centimeters = 1 hectometer.

To determine the relationship between meters and kilometers, start at the smallest unit and count up to kilometers by factors of 10.

$$
\begin{array}{ll}
1000 \rightarrow \text{km} & \text{since } 1000 \text{ m} = 1 \text{ km} \\
100 \rightarrow \text{hm} & \text{since } 100 \text{ m} = 1 \text{ hm} \\
10 \rightarrow \text{dam} & \text{since } 10 \text{ m} = 1 \text{ dam} \\
\text{Start} \rightarrow \text{m} &
\end{array}
$$

Thus 1000 meters = 1 kilometer

As with the American system, metric equivalent units may be converted to forms of 1 and used as factors in conversions. For example,

$$
\frac{1 \text{ km}}{1000 \text{ m}} = 1 \quad \text{or} \quad \frac{1000 \text{ m}}{1 \text{ km}} = 1
$$

$$
100,000 \frac{\text{cm}}{1 \text{ km}} = 1 \quad \text{or} \quad \frac{1 \text{ km}}{100,000 \text{ cm}} = 1
$$

When necessary, use the metric ladder to determine other relationships to use in unit conversions.

EXAMPLE 4   Perform the indicated unit conversions.

1. 34 dm = _____ mm   2. 250 m = _____ km

Solution   1. 34 dm = 34 dm $\dfrac{100 \text{ mm}}{1 \text{ dm}}$ Since 100 mm = 1 dm

$\qquad\qquad = 34 \cdot 100 \text{ mm}$ Rearranging products; dividing out $\dfrac{dm}{dm}$

$\qquad\qquad = 3400 \text{ mm}$

2. 250 m = 250 m $\dfrac{1 \text{ km}}{1000 \text{ m}}$ Since 1000 m = 1 km

$\qquad\quad = \dfrac{250}{1000} \text{ km}$ Rearranging products; dividing out $\dfrac{m}{m}$

$\qquad\quad = 0.25 \text{ km}$ ◄

Now that you have studied the conversions within the metric system, you should get a meter stick or ruler and examine the relative sizes of the millimeter, centimeter, decimeter, and meter. Measure things such as the width of your little finger and the width of your palm. Lay the meter stick on the floor and step off 1 meter. You will probably find that two normal walking steps is about 1 meter. Measure or step off 10 meters to visualize the length of 1 dekameter. Then measure 10 dekameters to see how long 1 hectometer is. The next time you take a walk, step off 1000 meters to determine the length of 1 kilometer.

QUICK CHECK

Perform the indicated conversions.

1. 2000 mm = _____ m   2. 6 km = _____ dam

3. 220 cm = _____ dm   4. 60 hm = _____ km

5. 0.04 km = _____ m

▶**3**   CONVERTING BETWEEN THE AMERICAN AND METRIC SYSTEMS

The connecting factors, called links, between the American and metric systems for measuring length are as follows.

LINKS BETWEEN THE AMERICAN AND METRIC SYSTEMS
1 inch = 2.54 centimeters
39.37 inches = 1 meter
1.09 yards = 1 meter

ANSWERS

1. 2   2. 600   3. 22   4. 6   5. 40

These decimal conversion factors, with the exception of centimeters to inches, are rounded to two decimal places, so they are approximations. Therefore, most unit conversions using these conversion factors will be approximations.

To change an American unit of measure to a metric unit or vice versa, you need only know one of the preceding links. However, some people prefer to learn several links to avoid extra steps in the conversion. The strategy for changing from the American system to the metric system or vise versa is to convert within the given system to get to the unit that links the system you are in to the system to which you are going. Then cross systems through the link conversion factor. Finally, convert within this system to obtain the desired unit.

EXAMPLE 5   Change 405 feet to decimeters.

Solution   First, change feet to inches. Next, change inches to centimeters using 2.54 cm = 1 in. Finally, change centimeters to decimeters.

$$405 \text{ ft} = 405 \frac{\text{ft}}{1} \cdot \frac{12 \text{ in.}}{1 \text{ ft}} \cdot \frac{2.54 \text{ cm}}{1 \text{ in.}} \cdot \frac{1 \text{ dm}}{10 \text{ cm}}$$

$$= \frac{405 \cdot 12 \cdot 2.54}{10}$$

$$= 1234.44 \text{ dm}$$

QUICK CHECK

Perform the indicated conversions.

1. 36 yd = _____ m   2. 15 cm = _____ ft

3. 8 ft = _____ dm   4. 0.001 in. = _____ mm

5. 5 ft = _____ dm

EXAMPLE 6   On the interstate highways of Texas the speed limit is 65 miles per hour. What is the speed limit in kilometers per hour?

Solution   Since both units are "per hour," we need only worry about changing 65 miles to kilometers.

$$65 \frac{\text{mi}}{\text{hr}} = 65 \frac{\text{mi}}{\text{hr}} \cdot \frac{5280 \text{ ft}}{1 \text{ mi}} \cdot \frac{12 \text{ in.}}{1 \text{ ft}} \cdot \frac{2.54 \text{ cm}}{1 \text{ in.}} \cdot \frac{1 \text{ km}}{100,000 \text{ cm}}$$

$$= \frac{65 \cdot 5280 \cdot 12 \cdot 2.54}{100,000}$$

$$= 105 \text{ km/hour (rounded to the units place)}$$

ANSWERS

1. 32.9   2. 0.49   3. 24.4   4. 0.0254   5. 15.24

A

ANSWERS

1. 2 mi = _____ yd

  1. _____

2. 600 ft = _____ in.

  2. _____

3. 100 yd = _____ mi

  3. _____

4. 1060 ft = _____ mi

  4. _____

5. 4 mi = _____ in.

  5. _____

6. 3 mi = _____ yd

  6. _____

7. 6 ft = _____ in.

  7. _____

8. 700 in. = _____ yd

  8. _____

9. 2,000,000 in. = _____ mi

  9. _____

10. 800 yd = _____ mi

  10. _____

11. 6000 m = _____ km

  11. _____

12. 265 dm = _____ cm

  12. _____

13. 35,000 mm = _____ dam

  13. _____

14. 4 hm = _____ m

  14. _____

15. 700 km = _____ dm

  15. _____

16. 40 mm = _____ dm

  16. _____

17. 500 cm = _____ mm

  17. _____

18. 2110 m = _____ hm

  18. _____

19. 6.5 hm = _____ km

  19. _____

20. 1.4 km = _____ m

  20. _____

21. 6.3 ft = _____ dam

  21. _____

22. 4.2 yd = _____ m

  22. _____

23. 65 hm = _____ mi

  23. _____

24. 40 cm = _____ in.

  24. _____

25. 6 ft = _____ dm

  25. _____

26. 32 ft = _____ dam

  26. _____

27. 75 in. = _____ dm

  27. _____

28. 3 mi = _____ km

  28. _____

29. 70 cm = _____ yd

  29. _____

30. 800 mm = _____ in.

  30. _____

31. Ann drives 5 miles to school each day. How far is this in kilometers?

  31. _____

32. It is 200 miles from Amador's house to his grandmother's house. How far is this in kilometers?

  32. _____

33. Al and Kay drive 8 miles a day taking their children to and from school. How far is this in hectometers?

33. _____

34. John and Julie drove 180 miles. How far is this in hectometers?

34. _____

35. Angenelle is 5 feet 4 inches tall. How tall is she in centimeters?

35. _____

36. Delores and Gus purchased a horse 17 hands tall (1 hand is about 4 inches). How tall is the horse in decimeters?

36. _____

37. Pat and Morris bought 30 yards of canvas to make a cover for their boat. How many meters is this?

37. _____

38. To make a dress, $3\frac{1}{4}$ yards of material are needed. How many meters of material are needed?

38. _____

39. Neil and Betty are considering buying some lakefront property with 200 feet of paved road access. How many meters is this?

39. _____

40. Leroy drove his tractor 62 kilometers while cutting a pasture. How far is this in miles?

40. _____

41. John's and Mary Alice's baby was 48.26 centimeters long at birth. How many inches is this?

41. _____

42. Bob and Cheryll are planning to put new carpet in their den. They measured the room and found that it was 10 meters by 12 meters. What does the room measure in feet?

42. _____

43. Marvin's hobby is long-distance bicycling. One weekend he rode 185 kilometers. How far is this in miles?

43. _____

44. Joe and Melinda went to Mexico on their vacation. They drove a total of 2000 kilometers. How far is this in miles?

44. _____

45. Bobby runs 5 kilometers a day. How far is this in miles?

45. _____

46. Patsy swam a 100-meter race. How far is this in yards?

46. _____

47. Martha's son B. J. won the 17-year-old bike-a-thon by riding 60 kilometers in one day. How far is this in feet?

47. _____

48. Janet purchased a new trailer that is 20 meters long. How long is the trailer in feet?

48. _____

49. Joyce strung 18 meters of popcorn to put on the Christmas tree. How many inches of popcorn did she string?

49. _____

50. Mavournee purchased 110 meters of canvas to make costumes for a play. How many yards of material did she buy?

50. _____

51. Mt. Everest is 29,028 feet tall at its highest peak. Find its height in miles.

51. _____

52. Lake Superior is 1333 feet deep at its deepest point. How many miles deep is it?

52. _____

53. The greatest depth of the Pacific Ocean is 35,760 feet. How deep is this in miles?

53. _____

54. The greatest depth of the Atlantic Ocean is 27,498 feet. How deep is this in miles?

54. _____

B

55. Billy walked 200 meters and then ran 1½ miles. How far did he travel in feet?

55. _____

56. Alex swam 100 meters one day and 600 yards the next. What is the total distance he swam in feet?

56. _____

57. When Shirley was seven years old she was 121.92 centimeters tall. By the time she was eight she was 4 feet 3 inches tall. How many inches had she grown?

57. _____

58. Dana grew 5 inches in one year to a height of 5 feet 4 inches. What was her height in centimeters at the beginning of the year?

58. _____

2.2 METRIC AND AMERICAN UNITS OF MASS

OBJECTIVE

Convert units of mass in the American system, the metric system, and between systems.

UNITS OF MASS

The basic units of measure for mass in the American System are the ounce, pound, and ton. Equivalent units and abbreviations for these units are as follows.

AMERICAN SYSTEM UNITS OF MASS	
Equivalent Units	Abbreviations
16 ounces = 1 pound	ounce = oz
2000 pounds = 1 ton	pound = lb

The fundamental unit for mass in the metric system is the gram. A paper clip has a mass of about one gram. Metric system units of mass are formed by using the same prefixes as those for length.

$$1 \text{ kilogram} = 1000 \text{ grams}$$
$$1 \text{ hectogram} = 100 \text{ grams}$$
$$1 \text{ dekagram} = 10 \text{ grams}$$
$$1 \text{ decigram} = \tfrac{1}{10} \text{ gram}$$
$$1 \text{ centigram} = \tfrac{1}{100} \text{ gram}$$
$$1 \text{ milligram} = \tfrac{1}{1000} \text{ gram}$$

Equivalent units and abbreviations for units of mass in the metric system are as follows.

METRIC SYSTEM UNITS OF MASS	
Equivalent Units	Abbreviations
1000 kilograms = 1 metric ton	kilogram = kg
10 hectograms = 1 kilogram	hectogram = hg
10 dekagrams = 1 hectogram	dekagram = dag
10 grams = 1dekagram	gram = g
10 decigrams = 1 gram	decigram = dg
10 centigrams = 1 decigram	centigram = cg
10 milligrams = 1 centigram	milligram = mg
Link Between American and Metric Systems	
2.2 pounds = 1 kilogram (approximately)	

The following metric ladder for units of mass may be used to convert from one unit to another.

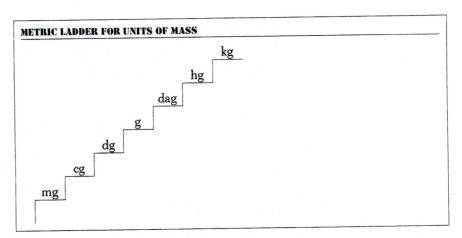

Notice that it takes a little more than 2 pounds to equal 1 kilogram. Thus 1 kilogram of butter has a mass of approximately 2.2 pounds.

Since the mass of a paper clip is about 1 gram, you can imagine how small 1 milligram is—it takes 1000 milligrams to equal to 1 gram.

We use the same strategy for converting units of mass as we did for converting units of length.

EXAMPLE 7   Perform the indicated conversions.

1. 1004 oz = _____ lb
2. How many metric tons does a 2-ton truck carry?
3. What is the mass in kilograms of a person whose mass is 185 pounds?
4. 2,000,506 mg = _____ hg
5. 53 dag = _____ cg

Solution

1. $1004 \text{ oz} = 1004 \text{ oz} \cdot \dfrac{1 \text{ lb}}{16 \text{ oz}}$

   $= \dfrac{1004}{16} \text{ lb}$

   $= 62.75 \text{ lb}$

2. $2 \text{ tons} = 2 \text{ tons} \cdot \dfrac{2000 \text{ lb}}{1 \text{ ton}}$

   $= 4000 \text{ lb} \cdot \dfrac{1 \text{ kg}}{2.2 \text{ lb}}$

   $= \dfrac{4000}{2.2} \text{ kg}$

   $= 1818.18 \text{ kg} \cdot \dfrac{1 \text{ metric ton}}{1000 \text{ kg}}$

   $= 1.818 \text{ metric tons}$

3. $185 \text{ lb} = 185 \text{ lb} \cdot \dfrac{1 \text{ kg}}{2.2 \text{ lb}}$

   $= \dfrac{185}{2.2} \text{ kg}$

   $= 84.09 \text{ kg}$

4. 100,000 ⟶ hg   since 100,000 mg = 1 hg
     10,000 ⟶ dag   since 10,000 mg = 1 dag
      1000 ⟶ g   since 1000 mg = 1 g
       100 ⟶ dg   since 100 mg = 1 dg
        10 ⟶ cg   since 10 mg = 1 cg
      Start ⟶ mg

   Thus 100,000 milligrams = 1 hectogram. Then

   $$2{,}000{,}506 \text{ mg} = 2{,}000{,}506 \text{ mg} \cdot \dfrac{1 \text{ hg}}{100{,}000 \text{ mg}}$$

   $$= 20.00506 \text{ hg}$$

5. $\begin{array}{ll} 1000 & \rightarrow \text{dag} \\ \quad 100 & \rightarrow \text{g} \\ \quad\quad 10 & \rightarrow \text{dg} \\ \text{Start} & \rightarrow \text{cg} \end{array}$ since 1000 cg = 1 dag
     since 100 cg = 1 g
     since 10 cg = 1 dg

Thus 1000 centigrams = 1 dekagram. Then

$$53 \text{ dag} = 53 \, \cancel{\text{dag}} \cdot \frac{1000 \text{ cg}}{1 \, \cancel{\text{dag}}}$$

$$= 53{,}000 \text{ cg}$$

◀

QUICK CHECK

Perform the indicated conversions.

1.   3 lb = _____ kg   2.   15 dg = _____ g

3.   150 lb = _____ dag   4.   36 oz = _____ lb

5.   1 ton = _____ metric ton

EXERCISE 2.2

<u>A</u>

ANSWERS

1.   200 lb = _____ ton

1. _____

2.   160 oz = _____ lb

2. _____

3.   256 lb = _____ oz

3. _____

4.   4 tons = _____ lb

4. _____

5.   10.6 lb = _____ oz

5. _____

6.   3000 lb = _____ tons

6. _____

7.   50 lb = _____ oz

7. _____

8.   17 oz = _____ lb

8. _____

9.   2563 oz = _____ lb

9. _____

10.   14 tons = _____ lb

10. _____

11.   4000 g = _____ hg

11. _____

12.   20 kg = _____ dg

12. _____

13.   15 dg = _____ mg

13. _____

14.   85 cg = _____ mg

14. _____

15.   2 metric tons = _____ hg

15. _____

16.   85 mg = _____ dg

16. _____

17.   5 hg = _____ g

17. _____

18.   25 dag = _____ dg

18. _____

19.   20 kg = _____ cg

19. _____

20.   25 kg = _____ g

20. _____

21.   110 lb = _____ kg

21. _____

22.   35 g = _____ oz

22. _____

23.   2600 hg = _____ lb

23. _____

24.   40 tons = _____ metric tons

24. _____

25.   2600 oz = _____ dg

25. _____

26.   3 lb = _____ g

26. _____

27.   40 oz = _____ g

27. _____

28.   1 oz = _____ g

28. _____

29.   2000 lb = _____ metric tons

29. _____

30.   1 ton = _____ metric tons

30. _____

31.   A turkey weighs 23 pounds. How much is this in kilograms?

31. _____

32.   Tom went on a diet and lost 15 pounds. How much weight in kilograms did he lose?

32. _____

33. What mass can a $1\frac{1}{2}$-ton truck carry in metric tons?

34. What is the mass in grams of $\frac{1}{2}$ pound of grapes?

35. A pharmacist mixed 140 milligrams of aspirin in a solution of cherry syrup. How many ounces of aspirin did he use?

36. A cup of water has a mass of 8 ounces. What is its mass in grams?

37. Barbara said that she wanted to gain 15 kilograms. How many pounds is this?

38. Reid bought a heifer at the cattle auction whose mass was 450 pounds. What was the heifer's mass in kilograms?

39. What is the mass of 30 grams of green beans in pounds?

40. What is the mass of 12,790 decigrams of potatoes in grams?

41. If the United States ever changes to the metric system, what will be the mass of a $1\frac{1}{4}$ ton truck's load in metric tons?

42. What is the mass in metric tons of a $\frac{3}{4}$-ton truck's load?

B

43. Mauri bought 3 pounds of ground beef to make a casserole. Her recipe called for 2 kilograms of ground beef. Did she buy enough?

44. If ground beef costs $1.69 per pound and Sylvia purchased 3 kilograms, how much did she pay for the meat?

45. Steel pipe is priced by mass per foot. A steel pipe 1 foot long has a mass of 17 pounds and costs $2.37. How much would a pipe of mass 5 kilograms cost?

46. Grass seed is priced by the pound and 1 pound costs $8. How much would 400 kilograms cost?

47. Sirloin steak costs $2.89 per pound. What is the cost per kilogram?

48. It takes 8 days for an ocean liner to cross the Atlantic. If it carries a crew of 753 and 1357 passengers, estimate how many pounds of meat the chef must stock for the voyage.

33. _____

34. _____

35. _____

36. _____

37. _____

38. _____

39. _____

40. _____

41. _____

42. _____

43. _____

44. _____

45. _____

46. _____

47. _____

48. _____

2.3 METRIC AND AMERICAN UNITS OF CAPACITY

OBJECTIVE

1 ▶ Convert units of capacity in the American system, the metric system, and between systems.

1 ▶ ## UNITS OF CAPACITY

Cola

3 liter

The fundamental units of capacity in the American system are the cup, pint, quart, and gallon. Equivalent units and abbreviations for these units of capacity are as follows.

AMERICAN SYSTEM UNITS OF CAPACITY	
Equivalent Units	Abbreviations
1 gallon = 4 quarts	gallon = gal
1 quart = 4 cups or 2 pints	quart = qt
1 pint = 2 cups	pint = pt
	cup = c

Since it is fairly common now for soft drinks to be sold by the 2 liters or 3 liters, Americans are beginning to "get the feel" for the fundamental metric unit of capacity.

The fundamental unit of capacity in the metric system is the liter. As with other metric units, the units of capacity are related by their prefixes.

$$1 \text{ kiloliter} = 1000 \text{ liters}$$
$$1 \text{ hectoliter} = 100 \text{ liters}$$
$$1 \text{ dekaliter} = 10 \text{ liters}$$
$$1 \text{ deciliter} = \frac{1}{10} \text{ liter}$$
$$1 \text{ centiliter} = \frac{1}{100} \text{ liter}$$
$$1 \text{ milliliter} = \frac{1}{1000} \text{ liter}$$

A liter is just slightly bigger than a quart (1 liter = 1.06 quarts). Since there are 4 quarts in a gallon, 4 liters is just slightly bigger than 1 gallon.

$$4 \text{ liters} = 4.24 \text{ quarts}$$
$$= 1.06 \text{ gallons}$$

The equivalent units and abbreviations for metric units of capacity are as follows.

METRIC SYSTEM UNITS OF CAPACITY

Equivalent Units	Abbreviations
10 milliters = 1 centiliter	kiloliter = kℓ
10 centiliters = 1 deciliter	hectoliter = hℓ
10 deciliters = 1 liter	dekaliter = daℓ
10 liters = 1 dekaliter	liter = ℓ
10 dekaliters = 1 hectoliter	deciliter = dℓ
10 hectoliters = 1 kiloliter	centiliter = cℓ
	milliliter = mℓ

Link Between American and Metric Systems

1 liter = 1.06 quarts (approximately)

As with length and mass, a metric ladder for units of capacity may be used to convert from one unit to another.

METRIC LADDER FOR UNITS OF CAPACITY

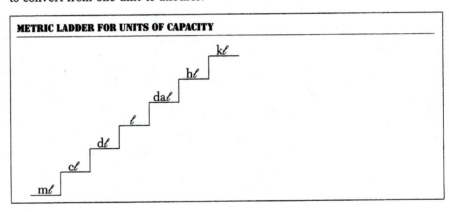

E X A M P L E 8   Perform the following conversions.

1. How many liters will a 16-gallon gasoline tank hold?
2. 1 cup = _____ hℓ?   3. 146 ℓ = _____ hℓ
4. 1 gal = _____ ℓ

Solution   1. $16 \text{ gal} = 16 \text{ gal} \cdot \dfrac{4 \text{ qt}}{1 \text{ gal}} \cdot \dfrac{1 \ell}{1.06 \text{ qt}}$ 2. $1 \text{ cup} = 1 \text{ cup} \cdot \dfrac{1 \text{ qt}}{4 \text{ c}} \cdot \dfrac{1 \ell}{1.06 \text{ qts}} \cdot \dfrac{10 \text{ d}\ell}{1 \ell}$

   $= 60.38 \ell$ $= 2.4 \text{ d}\ell$

3. $100 \overset{\frown}{} \text{h}\ell$ since 100 ℓ = hℓ
      $10 \overset{\frown}{} \text{da}\ell$ since 10 ℓ = 1 daℓ
   Start $\overset{\searrow}{} \ell$

   Thus 100 ℓ = 1 hℓ. Then

   $146 \ell = 146 \ell \cdot \dfrac{1 \text{ h}\ell}{100 \ell}$

   $= 1.46 \text{ h}\ell$

4. $1 \text{ gal} = 1 \text{ gal} \cdot \dfrac{4 \text{ qts}}{1 \text{ gal}} \cdot \dfrac{1 \ell}{1.06 \text{ qt}}$

   $= \dfrac{4}{1.06} \ell$

   $= 3.77 \ell$

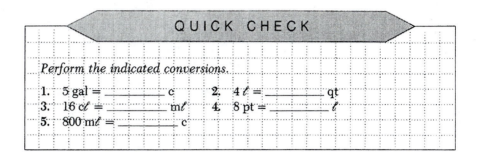

QUICK CHECK

Perform the indicated conversions.

1. 5 gal = _____ c
2. 4 ℓ = _____ qt
3. 16 cℓ = _____ mℓ
4. 8 pt = _____ ℓ
5. 800 mℓ = _____ c

ANSWERS

1. 80   2. 4.24   3. 160   4. 3.77   5. 3.4

EXERCISE 2.3

A

ANSWERS

Perform the indicated conversions.

1. 14 pt = _____ qt

2. 4 c = _____ gal

3. 5 gal = _____ qt

4. 16 gal = _____ pt

5. 40 qt = _____ gal

6. 50 gal = _____ ℓ

7. 3 qt = _____ c

8. 5 gal = _____ c

9. 5 c = _____ pt

10. 5 pt = _____ gal

11. 40 ℓ = _____ mℓ

12. 2067 cℓ = _____ dℓ

13. 96 dℓ = _____ hℓ

14. 40 daℓ = _____ mℓ

15. 22 kℓ = _____ cℓ

16. 200 mℓ = _____ ℓ

17. 3000 mℓ = _____ cℓ

18. 8 kℓ = _____ hℓ

19. 85 kℓ = _____ ℓ

20. 5 hℓ = _____ cℓ

21. 2 ℓ = _____ qt

22. 40 pt = _____ cℓ

23. 450 mℓ = _____ qt

24. 50 kℓ = _____ gal

25. 200 ℓ = _____ gal

26. 350 mℓ = _____ qt

27. 47 pt = _____ dℓ

28. 93 qt = _____ cℓ

29. 54 gal = _____ ℓ

30. 42 daℓ = _____ qt

31. How many quarts are in a 3-liter bottle of cola?

32. How many quarts are in a 2-liter bottle of cola?

1. _____

2. _____

3. _____

4. _____

5. _____

6. _____

7. _____

8. _____

9. _____

10. _____

11. _____

12. _____

13. _____

14. _____

15. _____

16. _____

17. _____

18. _____

19. _____

20. _____

21. _____

22. _____

23. _____

24. _____

25. _____

26. _____

27. _____

28. _____

29. _____

30. _____

31. _____

32. _____

33.   How many liters are in 2 gallons of milk?   33. _____

34.   How many liters are in 2 cups of milk?   34. _____

35.   A recipe calls for $2\frac{1}{2}$ cups of water. How many deciliters is this?   35. _____

36.   How many deciliters are there in 1 pint of ice cream?   36. _____

37.   If you drink 64 cups of water a day, how many liters of water do you   37. _____
      drink?

38.   A gasoline tank holds 50 liters. How many deciliters will it hold?   38. _____

39.   How many milliliters are there in 15 kiloliters of sulfuric acid?   39. _____

40.   How many centiliters are there in 19 hectoliters of water?   40. _____

B
—

41.   Gasoline is selling at $1.12 per gallon. How much does 1 liter cost?   41. _____

42.   One quart of milk costs $0.96. How much would 1 liter cost?   42. _____

43.   One-half gallon of orange juice costs $1.86. How much would 2 liters   43. _____
      cost?

44.   One-half gallon of chocolate milk costs $1.19. How much would 4 liters 44. _____
      cost?

45.   Fifty milliliters of nail polish cost $2.35. What is the price per liter?   45. _____

2.4 METRIC AND AMERICAN TEMPERATURE UNITS

OBJECTIVE

▶ Convert between degrees Celsius and degrees Fahrenheit.

▶ TEMPERATURE UNITS

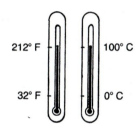

The American unit of measure for temperature is **degrees Fahrenheit**, °F. The metric unit for measuring temperature is **degrees Celsius**, °C. The metric temperature scale is determined from the freezing and boiling temperatures of water. In the metric system the freezing point of water is 0°C and the boiling point is 100°C. In the American system water freezes at 32°F and boils at 212°F.

There are two formulas used to convert between degrees Celsius and degrees Fahrenheit. To convert degrees Celsius to degrees Fahrenheit, substitute the given number of degrees Celsius for °C in the following formula and evaluate the resulting indicated operations.

$$°F = \frac{9}{5}°C + 32$$

or

$$°F = 1.8°C + 32 \quad \text{Since } \tfrac{9}{5} = 1.8$$

To convert degrees Fahrenheit to degrees Celsius, substitute the given number of degrees Fahrenheit for °F in one of the following formulas and evaluate the resulting indicated operations. Both formulas give the same result.

$$°C = \frac{5}{9}(°F - 32)$$

$$°C = \frac{°F - 32}{1.8} \quad \text{Note: } \tfrac{5}{9} \text{ is replaced with } \tfrac{1}{\tfrac{9}{5}} \text{ or } \tfrac{1}{1.8}$$

EXAMPLE 9

1. What would be the temperature setting in degrees Celsius for an oven temperature of 350°F?
2. Room temperature in the metric system is considered to be about 25°C. What is this temperature in degrees Fahrenheit?

Solution
1. $°C = \left(\dfrac{5}{9}\right)(°F - 32)$

$\quad = \left(\dfrac{5}{9}\right)(350 - 32) \quad$ Substituting 350 for °F

$\quad = \left(\dfrac{5}{9}\right)(318)$

$\quad = \dfrac{1590}{9}$

$\quad = 176\dfrac{2}{3}$

Thus $350°F = 176\tfrac{2}{3}°C$.

2. $°F = \left(\dfrac{9}{5}\right)°C + 32$

$\quad = \left(\dfrac{9}{5}\right)(25) + 32 \quad$ Substituting 25 for $°C$

$\quad = 45 + 32$

$\quad = 77$

Thus $25°C = 77°F$.

◀

QUICK CHECK

Perform the indicated temperature conversions.

1. $35°C =$ _____ $°F$.   2. $250°C =$ _____ $°F$.

3. $-40°F =$ _____ $°C$.   4. $75°F =$ _____ $°C$.

EXERCISE 2.4

A

ANSWERS

Perform the indicated temperature conversions.

1. 450° F = _____ ° C   1. _____

2. 90° F = _____ ° C   2. _____

3. 40° F = _____ ° C   3. _____

4. 46° F = _____ ° C   4. _____

5. 55° F = _____ ° C   5. _____

6. 95° F = _____ ° C   6. _____

7. 110° F = _____ ° C   7. _____

8. 99° F = _____ ° C   8. _____

9. 120° F = _____ ° C   9. _____

10. 85° F = _____ ° C   10. _____

11. 60° C = _____ ° F   11. _____

12. 15° C = _____ ° F   12. _____

13. 90° C = _____ ° F   13. _____

14. 5° C = _____ ° F   14. _____

15. 0° C = _____ ° F   15. _____

16. 85° C = _____ ° F   16. _____

17. 111° C = _____ ° F   17. _____

18. 119° C = _____ ° F   18. _____

19. 131° C = _____ ° F   19. _____

20. 136° C = _____ ° F   20. _____

21. Normal body temperature is 98.6° F. What is this temperature in degrees Celsius?   21. _____

22. A patient in the hospital has $2\frac{1}{2}°$ F of fever. How many degrees Celsius of fever is this?   22. _____

23. The thermostat in an office is set at 78° F. What is the setting in degrees Celsius?   23. _____

24. Cornbread is baked at an oven setting of 425° F. What is the setting in degrees Celsius?   24. _____

25. The normal temperature of a refrigerator is 40° F. What is this temperature in degrees Celsius?   25. _____

26. An outside temperature of 10° C is how many degrees Fahrenheit?   26. _____

27. Rocket boosters burn at 450° C. What is this temperature in degrees Fahrenheit?   27. _____

28. By mistake Robert set his oven to 350° C. His bread quickly burnt. What was the oven setting in degrees Fahrenheit?

28. _____

29. Coffee is generally served at about 75° C. What is this temperature in degrees Fahrenheit?

29. _____

30. During the summer, the ocean temperature is about 30° C. What is its temperature in degrees Fahrenheit?

30. _____

31. The sun's surface temperature is 10,000° F. What is this temperature in degrees Celsius?

31. _____

32. Sun spots are cooler than the surface of the sun, so they appear darker in color. The temperature of sun spots is about 2000° F less than the surface temperature of 10,000° F. What is the temperature in degrees Celsius of a sun spot?

32. _____

B

33. On June 30 at 6 A.M. the outside temperature was 73° F. At noon it was 93° F, and by 5 P.M. it was 87° F. What was this day's average temperature in degrees Celsius?

33. _____

34. In a laboratory experiment the temperature of a solution was read every 5 minutes for 30 minutes. The temperature readings were 21.8°, 27.9°, 28.3°, 30.7°, 35.3°, 31.7°, and 31.8°, all in degrees Celsius. What was the average temperature in degrees Fahrenheit?

34. _____

35. To cook fudge the mixture has to reach a temperature of 210° F. Once it reaches this temperature it must then cool to 100° F before adding butter and vanilla. What is this temperature difference in degrees Celsius?

35. _____

36. During the summer it costs $58.75 a month for each degree below 78° F that a thermostat in an office is set. What is the cost per degree Celsius?

36. _____

37. The cost for heating a house in the winter is $16.76 per month for each degree Fahrenheit above the outside temperature the house is heated. If the temperature of a house averaged 16° F above the outside temperature for one month, what will the heating bill be?

37. _____

38. When traveling abroad some people find that they can approximate a Fahrenheit temperature fairly accurately using the formula

$$°F = 2°C + 30$$

Use this formula to approximate the following Celsius temperatures, and then find the *actual* Fahrenheit temperatures using the formula

$$°F = \left(\frac{9}{5}\right)°C + 32$$

By how many degrees does this actual Fahrenheit temperature differ from the approximation? Do you see a difference in accuracy of the approximation for Celsius temperatures under 40° compared with those over 40° C?

a. 0° C   b. 15° C   c. 25° C   d. 35° C
e. 55° C   f. 70° C   g. 100° C

38. a. _____

    b. _____

    c. _____

    d. _____

    e. _____

    f. _____

    g. _____

2.5 FLUID MEASURES

O B J E C T I V E

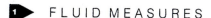

1 ▶ Convert fluid measures in the metric system, the American system, and between systems.

1 ▶ F L U I D   M E A S U R E S

You often see statements such as "16 fl oz" stamped on food products. The "fl" stands for fluid. In the American system 1 cup of water has a mass of 8 ounces. In the metric system 1 milliliter of water has a mass of 1 gram. It is customary to interchange these mass and capacity measures. For example, 6 grams is used to denote a capacity of 6 milliliters if the liquid has approximately the same mass as water. Similarly, when a recipe calls for 4 ounces of milk, you measure $\frac{1}{2}$ cup of milk. A measuring cup shows that 4 ounces is the same as $\frac{1}{2}$ cup. The assumption here is that milk or any other fluid has approximately the same mass as water. Thus, when you see "16 fl oz" you will know that the container has 2 cups of liquid and that the liquid is assumed to have a mass of 16 ounces (the mass of 2 cups of water).

It is not customary to precede a fluid or liquid metric measure with "fl." Even in the American system it is not always used. However, when you are asked to convert units of mass to units of capacity or vice versa, you should know that liquid measure is assumed.

E X A M P L E 1 0   1. 15 fl oz = _____ c   2. 22 g = _____ ℓ

3. 40 fl oz = _____ mℓ

Solution   1. $15 \text{ fl oz} = 15 \text{ fl oz} \cdot \dfrac{1 \text{ c}}{8 \text{ fl oz}}$

$= \dfrac{15}{8} \text{ c}$

$= 1.875 \text{ c}$

2. $22 \text{ g} = 22 \text{ g} \cdot \dfrac{1 \text{ m}\ell}{1 \text{ g}} \cdot \dfrac{1 \ell}{1000 \text{ m}\ell}$

$= \dfrac{22}{1000} \ell$

$= 0.022 \ell$

3. $40 \text{ fl oz} = 40 \text{ fl oz} \cdot \dfrac{1 \text{ c}}{8 \text{ fl oz}} \cdot \dfrac{1 \text{ qt}}{4 \text{ c}} \cdot \dfrac{1 \ell}{1.06 \text{ qt}} \cdot \dfrac{1000 \text{ m}\ell}{1 \ell}$

$= \dfrac{40 \cdot 1000}{8 \cdot 4 \cdot 1.06} \text{ m}\ell$

$= 1179.2 \text{ m}\ell$

QUICK CHECK

Perform the indicated conversions.

1. 6 c = _____ fl oz   2. 2 qt = _____ fl oz

3. 36 fl oz = _____ ℓ 4. 40 mg = _____ fl oz

5. 3 mℓ = _____ fl oz

ANSWERS

1. 48   2. 64   3. 1.06   4. 0.0014   5. 0.102

A

Perform the indicated conversions.

1. $\frac{1}{4}$ c = _____ fl oz

2. $\frac{2}{3}$ c = _____ fl oz

3. $1\frac{1}{2}$ c = _____ fl oz

4. $3\frac{1}{2}$ c = _____ fl oz

5. 1 pt = _____ fl oz

6. 2 pt = _____ fl oz

7. $1\frac{1}{2}$ pt = _____ fl oz

8. $3\frac{1}{4}$ pt = _____ fl oz

9. 2 qt = _____ fl oz

10. 3 qt = _____ fl oz

11. 1 ℓ = _____ g

12. 3 ℓ = _____ g

13. $2\frac{1}{2}$ ℓ = _____ g

14. $3\frac{1}{4}$ ℓ = _____ g

15. 200 mℓ = _____ g

16. 300 mℓ = _____ g

17. 150.8 mℓ = _____ g

18. 763.9 mℓ = _____ g

19. 90 cℓ = _____ mg

20. 80 cℓ = _____ mg

21. 30 fl oz = _____ c

22. 50 fl oz = _____ c

23. 75 fl oz = _____ pt

24. 90 fl oz = _____ pt

25. 250 fl oz = _____ qt

26. 300 fl oz = _____ qt

27. 600 g = _____ ℓ

28. 750 g = _____ ℓ

29. 2730 mg = _____ cℓ

30. 3540 mg = _____ cℓ

31. 21,765 cg = _____ mℓ

32. 14,475 cg = _____ mℓ

ANSWERS

1. _____

2. _____

3. _____

4. _____

5. _____

6. _____

7. _____

8. _____

9. _____

10. _____

11. _____

12. _____

13. _____

14. _____

15. _____

16. _____

17. _____

18. _____

19. _____

20. _____

21. _____

22. _____

23. _____

24. _____

25. _____

26. _____

27. _____

28. _____

29. _____

30. _____

31. _____

32. _____

33. 15 fl oz = _____ mℓ 33. _____

34. 25 fl oz = _____ mℓ 34. _____

35. 300 fl oz = _____ ℓ 35. _____

36. 750 fl oz = _____ ℓ 36. _____

37. 35 mℓ = _____ fl oz   37. _____

38. 75 mℓ = _____ fl oz   38. _____

39. 20 ℓ = _____ fl oz   39. _____

40. 35 ℓ = _____ fl oz   40. _____

B

41. A company prices its grapefruit juice by the fluid ounce. One fluid ounce   41. _____
     costs $0.04. How much would the company charge for a 16 fl oz container
     of juice?

42. Orange juice is priced at $0.05 per fluid ounce. What would be the price   42. _____
     of $\frac{1}{2}$ gallon of juice?

43. Tomato juice costs $0.08 per fluid ounce. What is the price of 2 cups of   43. _____
     juice?

44. A 2-liter bottle of cola costs $1.89. What is the price per fluid ounce?   44. _____

45. A 3-liter bottle of cola costs $2.65. What is the cost per fluid ounce?   45. _____

Perform the following unit conversions:

ANSWERS

1. 60 ft = _____ yd

2. 40 yd = _____ in.

3. 73 in. = _____ ft

4. 3 mi = _____ ft

5. 15 m = _____ hm

6. 200 cm = _____ dm

7. 3000 mm = _____ m

8. 500 dam = _____ km

9. 23 in. = _____ cm

10. 65 m = _____ yd

11. 65 km = _____ mi

12. 900 km = _____ ft

13. 40 lb = _____ oz

14. 16 tons = _____ lb

15. 1.4 kg = _____ g

16. 1.9 dg = _____ hg

17. 14 kg = _____ lb

18. 26 oz = _____ dg

19. 20 tons = _____ metric tons

20. 5 gal = _____ pt

21. 6 c = _____ qt

22. 8 cℓ = _____ ℓ

23. 14 mℓ = _____ dℓ

24. 85 hℓ = _____ kℓ

25. 40 daℓ = _____ kℓ

26. 26 gal = _____ ℓ

27. 14 pt = _____ dℓ

28. 85 ℓ = _____ qt

29. 16 gal = _____ hℓ

30. 85° F = _____ °C

31. 36° F = _____ °C

32. 40° C = _____ °F

33. 85° C = _____ °F

34. 15 c = _____ fl oz

1. _____

2. _____

3. _____

4. _____

5. _____

6. _____

7. _____

8. _____

9. _____

10. _____

11. _____

12. _____

13. _____

14. _____

15. _____

16. _____

17. _____

18. _____

19. _____

20. _____

21. _____

22. _____

23. _____

24. _____

25. _____

26. _____

27. _____

28. _____

29. _____

30. _____

31. _____

32. _____

33. _____

34. _____

35.   1 pt = _____ fl oz   35. _____

36.   3 qt = _____ fl oz   36. _____

37.   42 mℓ = _____ g of water   37. _____

38.   24 ℓ = _____ kg of water   38. _____

Perform each unit conversion.

1.   43 ft = _____ yd

2.   79 in. = _____ ft

3.   12 cm = _____ m.

4.   63 m = _____ yd

5.   2759 ft = _____ mi

6.   70 mi = _____ km

7.   200 km = _____ dm

8.   12 lb = _____ oz

9.   16 kg = _____ lb

10.   29 kg = _____ dag

11.   12 tons = _____ metric tons

12.   17 c = _____ qt

13.   14 gal = _____ ℓ

14.   250 mℓ = _____ dℓ

15.   35° F = _____ °C

16.   76° C = _____ °F

17.   23 fl oz = _____ c

18.   3 qt = _____ fl oz

1. _____

2. _____

3. _____

4. _____

5. _____

6. _____

7. _____

8. _____

9. _____

10. _____

11. _____

12. _____

13. _____

14. _____

15. _____

16. _____

17. _____

18. _____

GEOMETRICAL FIGURES

3.1 POINTS, LINES, PLANES, AND SPACE

OBJECTIVES

1 ▶ Characterize and symbolically represent points, lines, planes, and space.

2 ▶ State the requirements for determining points, lines, planes, and space; define the terms coplanar and collinear; classify planes by their relative position.

3 ▶ Classify lines by their relative position; show how two intersecting lines determine a plane.

4 ▶ Define the term "between."

5 ▶ Define and symbolically represent a line segment and its measure; state the conditions for geometrical sets to be congruent.

6 ▶ Define the terms bisect, midpoint, ray, half-line, half-plane, and half-space.

7 ▶ Describe perpendicular and parallel lines and planes; explain the relative positions for a line and a plane.

Geometry deals primarily with sets of points. Familiar geometrical figures such as lines, triangles, and circles are simply particular types of point sets. In this section we begin our study of geometry by developing a vocabulary. There are many basic concepts discussed in this section that you should study carefully and become familiar with. Most of these concepts are represented by a drawing. You will find it helpful to study these illustrations to develop mental pictures that reinforce the concepts.

In geometry there are several fundamental undefined terms: a point, a line, a plane, and space. We begin our study of geometry by discussing these undefined terms. Since these terms are undefined, we do not attempt to define them; instead we discuss how they are represented symbolically and point out some of their characteristics.

1 ▶ POINTS, LINES, PLANES, AND SPACE

Points in geometry are designated by dots and are usually named with capital printed letters. Figure 3.1 shows three points designated by the letters A, B, and C. Points are without dimension, which means that they have no length, width, or height.

A **line** is a set of points. The line in Figure 3.2 is designated by naming two points on the line and writing $\overleftrightarrow{AB}$ or by writing "line l." In either case the designation refers to a set of points. $\overleftrightarrow{AB}$ is read "line AB." Lines have the dimension of length only.

A **plane** is a set of points that forms a flat surface. Figure 3.3 shows that planes are designated by a single capital script letter placed in a corner. In Figure 3.3 we refer to the plane as "plane $\mathcal{M}$." Planes have two dimensions; that is, they have no thickness but extend indefinitely in two dimensions, length and width.

Space is the set of all points and has no special designation. Space extends indefinitely in three dimensions—length, width, and height. We refer to a drawing

• A

    • B

• C

FIGURE 3.1   Points A, B, and C

FIGURE 3.2   Line AB, $\overleftrightarrow{AB}$, or line l

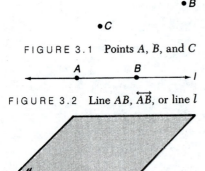

FIGURE 3.3   Plane $\mathcal{M}$

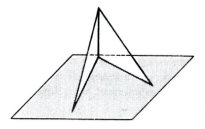

FIGURE 3.4
Space figure

such as that in Figure 3.4 as a space figure or a three-dimensional figure; we can also say that it occupies space or that it is in three-space.

QUICK CHECK

Answer each of the following true or false.

1. A point is represented by a dot.
2. Planes have thickness.
3. The set of all points is space.
4. A line contains at least two points.
5. Point, line, plane, and space are defined terms.

▶2 DETERMINING POINTS, LINES, PLANES, AND SPACE; RELATIVE POSITIONS OF LINES AND PLANES

FIGURE 3.5
Two points determine a line

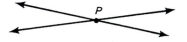

FIGURE 3.6
Two intersecting lines determine a point

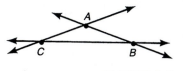

FIGURE 3.7
Noncollinear points *A*, *B*, and *C*

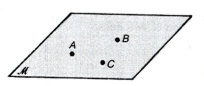

FIGURE 3.8
A plane is determined by three · noncollinear points

Now that we have discussed the basic undefined terms, let's consider some ways that these sets can be determined. Unless otherwise instructed, when we refer to two lines, two points, and so on, we mean **distinct** sets. The word **determine** describes the minimum conditions that guarantee the existence of a geometrical point set. When we say "two points determine a line," we mean that if you are given two points, then there exists exactly one line that contains the two points (Figure 3.5).

When two lines intersect, their intersection is a point. Therefore, two intersecting lines determine a point—their point of intersection (Figure 3.6).

If several points are contained in the same line, they are collinear. The word **collinear** means "belonging to the same line." Thus, if *A* and *B* are two points, they must be collinear. Three points, however, may or may not be collinear. **Noncollinear** means that there is no line that will contain a given set of points. The set of points *A*, *B*, and *C* represented in Figure 3.7 are noncollinear because there is no one line that contains them all. However, any two of the points are collinear because there is a line that contains them. Remember, two distinct points determine a line.

If you have three noncollinear points, then there is exactly one plane that will contain them. Therefore, three noncollinear points determine a plane. In Figure 3.8, points *A*, *B*, and *C* are noncollinear points and there is exactly one plane that contains them.

If a set of points can be contained by a plane, then the set is **coplanar**. Any set of two points is coplanar. As you can see from Figure 3.9 (on the next page), a set of two points may be contained by more than one plane. However, for them to be coplanar they only need to be contained in at least one plane.

ANSWERS

1. True 2. False 3. True 4. True 5. False

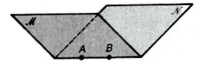

FIGURE 3.9
Two points are coplanar

FIGURE 3.10
Classification of planes by relative position

Any set of three points is coplanar. If the points are collinear, then there is more than one plane that contains them; if they are noncollinear, then there is exactly one plane that contains them. In either case they are coplanar. **Noncoplanar** means that there is no plane that contains a given set of points.

Two distinct planes either intersect or they do not intersect. If two planes do not intersect, they are **parallel planes**. If two planes intersect, then their intersection is a line. See Figure 3.10.

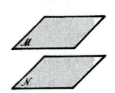

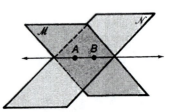

(a) Parallel planes $\mathcal{M}$ and $\mathcal{N}$

(b) Intersecting planes
$\mathcal{M} \cap \mathcal{N} = \overleftrightarrow{AB}$

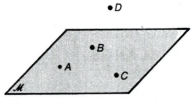

FIGURE 3.11
Four noncoplanar points determine space

Four noncoplanar points determine space. In Figure 3.11 points A, B, and C are coplanar. However, there is no single plane that contains points A, B, C, and D; therefore they are noncoplanar. Since these points cannot be contained in a plane, they must lie in space. We say that they "determine space."

To summarize,

1. Two intersecting lines determine a point.
2. Two points determine a line.
3. Three noncollinear points determine a plane.
4. Four noncoplanar points determine space.

QUICK CHECK

Answer each of the following true or false.

1. Two points are always collinear as well as coplanar.
2. The intersection of two planes is a line.
3. If four points are noncoplanar, then they lie in space.
4. Three points determine a line.
5. Three points can be coplanar.
6. Two points determine a line.
7. Two planes can intersect in a point.
8. Planes that do not intersect are called parallel planes.
9. A point is determined by the intersection of two lines.
10. Two distinct planes can be coplanar.

ANSWERS

1. True 2. True 3. True 4. False 5. True 6. True 7. False 8. True 9. True
10. False

3 RELATIVE POSITION OF LINES; PLANES DETERMINED BY INTERSECTING LINES

Now let's examine the relative positions for two lines. Given two lines, they are either coplanar or noncoplanar. If they are noncoplanar, they are called **skew**. If two distinct lines are coplanar, they either intersect in exactly one point and are called **intersecting lines** or they do not intersect and are called **parallel lines**. The phrase "is parallel to" is represented symbolically by ‖. In Figure 3.12, $\overleftrightarrow{AB}$ and $\overleftrightarrow{CD}$ are skew lines, $\overleftrightarrow{EF}$ and $\overleftrightarrow{GH}$ are intersecting lines, and $\overleftrightarrow{IJ} \parallel \overleftrightarrow{KL}$. Figure 3.12 shows all possible relative positions for two lines.

FIGURE 3.12
Classification of lines by relative positions

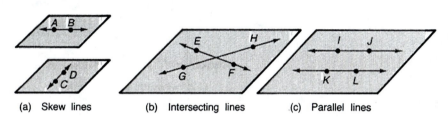

(a) Skew lines   (b) Intersecting lines   (c) Parallel lines

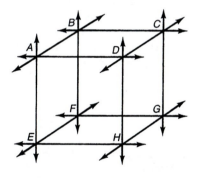

FIGURE 3.13
Parallel, skew, and intersecting lines

If two lines do not intersect, are they parallel? Not necessarily—they could be skew. In Figure 3.13 there are several pairs of skew, parallel, and intersecting lines. One pair of skew lines is $\overleftrightarrow{AB}$ and $\overleftrightarrow{DH}$. Can you name others? Are $\overleftrightarrow{EF}$ and $\overleftrightarrow{DC}$ skew? Lines $\overleftrightarrow{AB}$ and $\overleftrightarrow{HG}$ are parallel, as are $\overleftrightarrow{AB}$ and $\overleftrightarrow{EF}$. Look for other parallel lines. There are many pairs of intersecting lines, such as $\overleftrightarrow{EH}$ and $\overleftrightarrow{HG}$. How many pairs of intersecting lines are shown?

The two intersecting lines shown in Figure 3.14 must contain at least three noncollinear points or they would not be distinct. Therefore, two intersecting lines guarantee a minimum of three noncollinear points. And, of course, three noncollinear points determine a plane. Thus, two intersecting lines determine a plane because they guarantee three noncollinear points.

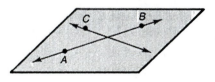

FIGURE 3.14
Two intersecting lines determine a plane

QUICK CHECK

Answer each of the following true or false.

1. Intersecting lines determine a point as well as a plane.

2. Skew lines are noncoplanar.

3. Parallel lines are coplanar.

4. Parallel lines as well as skew lines do not intersect.

5. ‖ is the symbol for "is parallel to."

ANSWERS

1. True   2. True   3. True   4. True   5. True

▶4   DEFINITION OF ''BETWEEN''

• B

• A   • C

FIGURE 3.15
B is not between A and C

One of the most important concepts in geometry is the concept of ''betweeness.'' A point B is **between** points A and C if the distance from A to B plus the distance from B to C is equal to the distance from A to C. In Figure 3.15, is B between A and C? No, because the distance from A to B plus the distance from B to C is not equal to the distance from A to C. What must be true of A, B, and C for B to be between A and C? They must be collinear. In Figure 3.16, R is between P and Q because the distance from P to R plus the distance from R to Q is equal to the distance from P to Q.

```
      P       R           Q
◄─────●───────●───────────●─────►
```

FIGURE 3.16
R is between P and Q

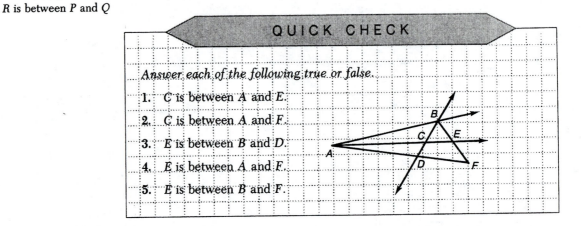

QUICK CHECK

Answer each of the following true or false.

1. C is between A and E.
2. C is between A and F.
3. E is between B and D.
4. E is between A and F.
5. E is between B and F.

▶5   LINE SEGMENTS AND THEIR MEASURE; CONGRUENCY

If P and Q are points, then the set of all collinear points between P and Q, including P and Q, is called **line segment PQ**. The set of points making up line segment PQ is designated $\overline{PQ}$. Points P and Q are called the **endpoints** of the line segment. To represent a line segment without its endpoints we write $\overset{\circ\circ}{PQ}$, and to represent a line segment minus one endpoint, we write either $\overset{\circ}{PQ}$ or $\overline{PQ}\!\!{}^\circ$. The set of points between P and Q, but not including P and Q, is called the **interior of $\overline{PQ}$**. Thus $\overset{\circ\circ}{PQ}$ is the interior of $\overline{PQ}$. See Figure 3.17.

FIGURE 3.17
Line segments without endpoints

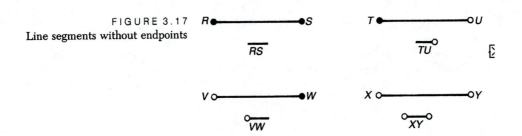

The **measure** or **length** of line segment $\overline{PQ}$ is designated by PQ and can be stated in inches, meters, and so on, or in unspecified units. The symbol PQ represents the measure (a number) of line segment $\overline{PQ}$, whereas the symbol $\overline{PQ}$ represents the set of points between P and Q. Consequently, $\overline{PQ}$ and PQ are never equal because PQ represents a number and $\overline{PQ}$ represents a set of points. For example, if a line segment PR is 3 inches long, we write $PR = 3$ inches, but we would never write $\overline{PR} = 3$ inches.

Two distinct geometrical sets of points that have the same size and shape are not equal because they are different sets; however, they do have common characteristics. Two line segments both 3 inches long are not equal, but they do have equal measures. Similarly, two circles may be the exact same size, but they are not equal because they are different sets. If two geometric figures have exactly the same size and shape, they are **congruent**. The symbol $\cong$ means "is congruent to." Throughout our study of geometry we will examine conditions that necessitate two sets of points being congruent. The first congruent sets we look at are congruent line segments.

Two line segments are congruent if and only if they have the same measure. In symbols, $\overline{AB} \cong \overline{PQ}$ iff $AB = PQ$. Just because $\overline{AB} \cong \overline{PQ}$ does not mean that $\overline{AB}$ and $\overline{PQ}$ are necessarily the same sets of points; they just have the same measure. In Figure 3.18, line segments PQ and AB are congruent because they have the same measure. $\overline{PQ}$ and $\overline{CD}$ are not congruent because their measures are different.

FIGURE 3.18
$\overline{AB}$ and $\overline{PQ}$ are congruent

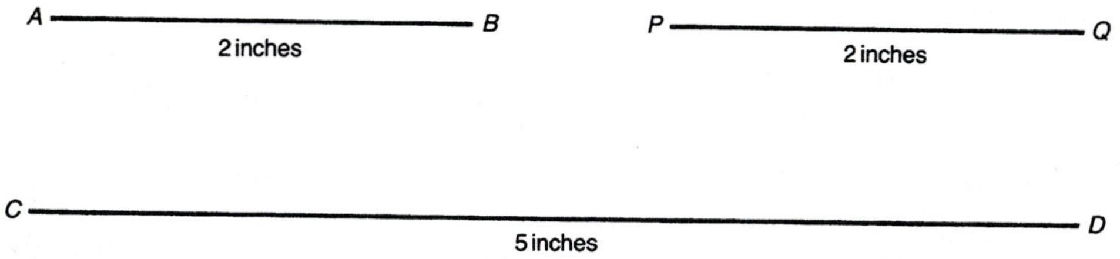

A ———————————————— B   P ———————————————— Q
        2 inches   2 inches

C ——————————————————————————————————— D
                        5 inches

QUICK CHECK

Answer each of the following true or false.

1. $\overset{\circ\circ}{XY}$ represents the interior of $\overline{XY}$.

2. $XY \neq \overline{XY}$.

3. If $\overline{AB} \cong \overline{CD}$, then they are the same line segment.

4. If $RS = VT$, then RS and VT have the same measure.

5. Congruent line segments have equal measures.

ANSWERS

1. True **2.** True **3.** False **4.** True **5.** True

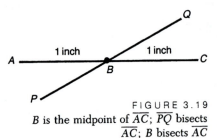

FIGURE 3.19
B is the midpoint of $\overline{AC}$; $\overleftrightarrow{PQ}$ bisects $\overline{AC}$; B bisects $\overline{AC}$

▶ 6 BISECTION, RAYS, HALF-PLANES, AND HALF-SPACE

The concept of dividing a geometrical figure into two halves is one that is repeatly considered in geometry. The point that divides a line segment into two equal halves is called its **midpoint**. If point B is between points A and C, and if $AB = BC$, then B is called the midpoint of $\overline{AC}$ and is said to bisect $\overline{AC}$. Any geometrical set other than the given line segment itself that contains the midpoint of the given line segment is said to **bisect** the given line segment. In Figure 3.19, the line segment PQ bisects line segment AC because $\overleftrightarrow{PQ}$ contains the midpoint of $\overline{AC}$.

Now let's look at several concepts that are sometimes called the **rules of separation**. In geometry certain geometrical figures separate certain other geometrical figures into three disjoint subsets. A point divides a line into three sets, a line divides a plane into three sets, and a plane divides space into three sets. As these separations occur, new geometrical figures are determined and named.

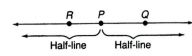

FIGURE 3.20
A point separates a line into three disjoint sets

RULES OF SEPARATION FOR A POINT AND A LINE

A point separates a line into three disjoint sets:

1. The set containing the point
2. The set of points on one side of the point
3. The set of points on the other side of the point

Each of the sets described in (2) and (3) are called **half-lines**. The point described in (1) is called the **endpoint of the half-line**, although the half-line itself does not contain its endpoint. A half-line is sometimes called the **side of a point**. A half-line is designated by naming its endpoint and one point on the half-line. In Figure 3.20, the half-line to the right of P is designated $\overset{\circ}{PQ}$, and the half-line to the left of P is designated $\overset{\circ}{PR}$.

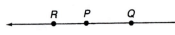

FIGURE 3.21
Three collinear points determine several half-lines

In Figure 3.21, we can name the following half-lines: $\overset{\circ}{PQ}$, $\overset{\circ}{PR}$, $\overset{\circ}{QP}$, $\overset{\circ}{QR}$, $\overset{\circ}{RP}$, and $\overset{\circ}{RQ}$. Notice that $\overset{\circ}{RP}$ and $\overset{\circ}{RQ}$ name the same half-line and that $\overset{\circ}{QP}$ and $\overset{\circ}{QR}$ also name the same half-line. Therefore we can say that $\overset{\circ}{RP} = \overset{\circ}{RQ}$ and that $\overset{\circ}{QP} = \overset{\circ}{QR}$. Remember two things: (1) a half-line is a set of points, and (2) two sets are equal only when they have exactly the same elements.

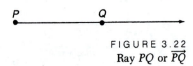

FIGURE 3.22
Ray PQ or $\overrightarrow{PQ}$

The union of a half-line with the set containing its endpoint is called a **ray**. We could say that a ray is a half-line united with its endpoint. A ray is named by naming its endpoint first and then one other point on the ray. A ray is represented symbolically as $\overrightarrow{PQ}$. See Figure 3.22.

Two different rays that have a common endpoint and are collinear are called **opposite rays**. See Figure 3.23.

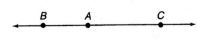

FIGURE 3.23
$\overrightarrow{AB}$ and $\overrightarrow{AC}$ are opposite rays

Line segments, rays, and half-lines are considered *parallel* if they are subsets of parallel lines and *skew* if they are subsets of skew lines. In Figure 3.24 $\overrightarrow{AB}$ and $\overrightarrow{CD}$ are skew rays because they are subsets of skew lines. Furthermore, $\overline{EC}$ and $\overline{AB}$ are parallel line segments because they are subsets of parallel lines.

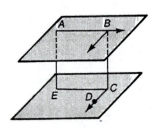

FIGURE 3.24
Skew and parallel rays and line segments

RULES OF SEPARATION FOR A LINE AND A PLANE

A line separates a plane into three disjoint sets:

1. The set of points on the line
2. The set of points on one side of the line
3. The set of points on the other side of the line

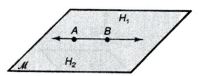

FIGURE 3.25
Half-planes

The two sets described in (2) and (3) are called **half-planes**. In Figure 3.25 $\overleftrightarrow{AB}$ divides plane $\mathcal{M}$ into three disjoint sets: half-plane H_1, half-plane H_2, and $\overleftrightarrow{AB}$. The half-planes are called **sides of the line**, and $\overleftrightarrow{AB}$ is called the **edge** of the half-planes. Note that a half-plane does not contain its edge.

RULES OF SEPARATION FOR A PLANE AND SPACE

A plane divides space into three disjoint sets:

1. The set of points in the plane
2. The set of points on one side of the plane
3. The set of points on the other side of the plane

The two sets described in (2) and (3) are called **half-spaces** or **sides of the plane**. In Figure 3.26 plane $\mathcal{M}$ divides space into three disjoint sets: plane $\mathcal{M}$, half-space H_1, and half-space H_2.

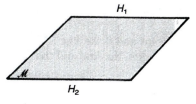

FIGURE 3.26
Half-spaces

QUICK CHECK

Answer each of the following true or false.

1. A half-plane contains its edge.

2. Half-planes are also called sides of a line.

3. A half-line contains its endpoint.

4. A ray is the union of a half-line and its endpoint.

5. $\overleftrightarrow{PQ}$ represents ray PQ.

7 PERPENDICULAR AND PARALLEL
LINES AND PLANES

Thus far we have discussed relative positions for two lines and two planes. Similarly, we could discuss the relative positions of line segments, rays, half-lines, half-planes, and various combinations of these. However, it is left to the student to consider the conditions under which rays, line segments, half-lines, and so on are parallel, intersect, or are skew. There is only one other combination that we examine here—the relative positions for a line and a plane. In this discussion we introduce the concept of perpendicular point sets. In retrospect, you should be able to apply this concept to rays, half-lines, line segments, and so on.

Two lines are perpendicular if and only if they intersect forming right angles. (As you will see in Section 3.2, right angles are angles whose measure is 90 degrees.) In Figure 3.27, $\overleftrightarrow{AB}$ is perpendicular to $\overleftrightarrow{CD}$. (Note that the square in the corner in Figure 3.27 indicates a right angle.) The phrase "is perpendicular to" is represented by the symbol $\perp$. Thus, $\overleftrightarrow{AB} \perp \overleftrightarrow{CD}$ is read "line AB is perpendicular to line CD."

If a line and a plane have no points in common, they are **parallel**. If a line and a plane intersect, then they have exactly one point in common or the line

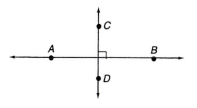

FIGURE 3.27
$\overleftrightarrow{AB}$ is perpendicular to $\overleftrightarrow{CD}$

FIGURE 3.28
Relative positions for a line and
a plane

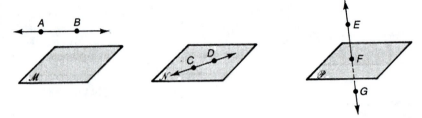

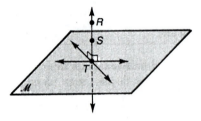

FIGURE 3.29
$\overleftrightarrow{RS} \perp$ plane $\mathcal{M}$

lies in the plane. If the intersection of a line and a plane is exactly one point, then the point is called the **foot of the line**. In Figure 3.28, $\overleftrightarrow{AB} \parallel$ plane $\mathcal{M}$; $\overleftrightarrow{CD} \in$ plane $\mathcal{N}$; $\overleftrightarrow{CD} \cap$ plane $\mathcal{N} = \overleftrightarrow{CD}$; $\overleftrightarrow{EG} \cap$ plane $\mathcal{P} = \{F\}$; and F is the foot of $\overleftrightarrow{EG}$.

If a line and a plane intersect in exactly one point and if the line is perpendicular to every line in the plane containing its foot, then the line and the plane are perpendicular. In Figure 3.29, $\overleftrightarrow{RS} \perp$ plane $\mathcal{M}$.

QUICK CHECK

Answer each of the following true or false.

1. A plane and a line can be both parallel and perpendicular.

2. A line can never intersect a plane in more than one point.

3. If two points on line l lie in plane $\mathcal{M}$, then line $l \subseteq \mathcal{M}$.

4. In general, "perpendicular" means that two geometrical sets intersect in such a way that right angles are formed.

5. A line and a plane that do not intersect are called skew.

EXAMPLE 1   Refer to the figure shown and find the following intersections or unions.

1. $\overleftrightarrow{AB} \cap \overleftrightarrow{AH}$
2. $\overleftrightarrow{KD} \cup \overleftrightarrow{EF}$
3. $\overrightarrow{DA} \cap \overrightarrow{BA}$
4. $\overline{DE} \cup \overline{EF}$
5. $\overline{DE} \cap \overline{DK}$
6. $\overrightarrow{EA} \cup \{E\}$

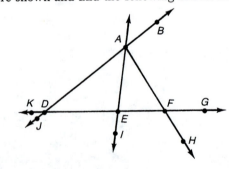

Solution   1. $\overleftrightarrow{AB} \cap \overleftrightarrow{AH} = \{A\}$ 2. $\overleftrightarrow{KD} \cup \overleftrightarrow{EF} = \overleftrightarrow{EF}$
3. $\overrightarrow{DA} \cap \overrightarrow{BA} = \overline{DB}$ 4. $\overline{DE} \cup \overline{EF} = \overline{DF}$
5. $\overline{DE} \cap \overline{DK} = \phi$ 6. $\overrightarrow{EA} \cup \{E\} = \overrightarrow{EA}$

ANSWERS

1. False   2. False   3. True   4. True   5. False

◄ E X E R C I S E 3 . 1

A

ANSWERS

Answer each of the following true or false. If your answer is false, explain why.

1. A plane must contain at least two points.

2. A plane must contain at least three points.

3. Any two points are collinear.

4. Any three points are coplanar.

5. If two planes intersect, then their intersection can be a point.

6. If two planes intersect, then their intersection can be a line segment.

7. $\overrightarrow{AB} = \overrightarrow{BA}$

8. $\overset{\circ}{\overrightarrow{AB}} = \overset{\circ}{\overrightarrow{BA}}$

9. $\overline{AB} = \overline{BA}$

10. $AB = BA$

11. By the definition of "between," P must be between P and Q.

12. If A, B, and C are collinear points, then $\overrightarrow{AB} = \overrightarrow{AC}$.

13. If two lines do not intersect, then they are parallel lines.

14. Intersecting lines are coplanar.

15. Every line is the subset of some plane.

16. Skew lines are coplanar.

17. Three collinear points determine a plane.

18. $\overline{AB} \cap \overline{BC} = \{B\}$

19. A ray contains its endpoint.

20. A half-line contains its endpoint.

21. Refer to the figure shown and name a pair of intersecting planes, a pair of perpendicular planes, and a pair of parallel planes.

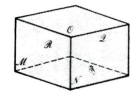

1. _____

2. _____

3. _____

4. _____

5. _____

6. _____

7. _____

8. _____

9. _____

10. _____

11. _____

12. _____

13. _____

14. _____

15. _____

16. _____

17. _____

18. _____

19. _____

20. _____

21. _____

Refer to the figure shown and find each of the following intersections or unions.

22. $\overleftrightarrow{AB} \cap \overrightarrow{BC}$

23. $\overrightarrow{AC} \cap \overrightarrow{CA}$

24. $\overleftrightarrow{AB} \cap \overrightarrow{AC}$

25. $\overset{\circ}{\overrightarrow{AB}} \cup \{A\}$

26. $\overrightarrow{CA} \cup \overset{\circ}{\overleftrightarrow{CD}}$.

27. $\overline{CB} \cap \overset{\circ}{\overrightarrow{AB}}$

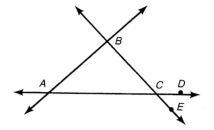

22. _____

23. _____

24. _____

25. _____

26. _____

27. _____

28. $\overrightarrow{AD} \cap \overleftrightarrow{CD}$ 28. _____

29. $\overleftrightarrow{BC} \cup \overrightarrow{BE}$ 29. _____

30. $\overleftrightarrow{AC} \cup \overleftrightarrow{CD}$ 30. _____

Make a drawing for Exercises 31–38.

31. Two planes that intersect in one line   31. _____

32. Three lines, any two of which are skew   32. _____

33. Two line segments that are collinear but have no intersection   33. _____

34. Four points located so that they determine exactly four lines   34. _____

35. Four points located so that they determine exactly six lines   35. _____

36. Four points located so that they determine exactly one line   36. _____

37. A line segment that is bisected by a ray   37. _____

38. Opposite rays that intersect   38. _____

B

39. Use a geometrical concept to explain why some four-legged chairs rock.   39. _____

40. Why do you think a tripod is used to support a camera?   40. _____

41. How many lines do five noncollinear points determine?   41. _____

42. How many lines do 200 noncollinear points determine?   42. _____

3.2 ANGLES AND TRIANGLES

OBJECTIVES

1 ▶ Define an angle, represent it symbolically, and explain how it is determined and measured; add and subtract angle measures.

2 ▶ Define congruent angles; state the Angle Addition Property; define acute, obtuse, and right angles; identify perpendicular lines and perpendicular subsets of a line.

3 ▶ Identify adjacent and vertical angles.

4 ▶ Identify and name angles formed by parallel lines cut by a transversal; state which of these angles are congruent and which are supplementary.

5 ▶ Define an angle bisector; identify and name a triangle and its parts.

6 ▶ Classify triangles by relative side lengths, relative angle measures, and combinations of these.

7 ▶ Identify adjacent and opposite sides and angles of a triangle; identify the base and vertex angles of an isosceles triangle.

8 ▶ Define medians and altitudes of a triangle.

1 ▶ ANGLES AND THEIR MEASURE

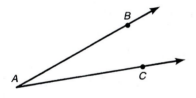

FIGURE 3.30
An angle is the union of two rays with a common endpoint

An **angle** is the union of two rays with a common endpoint. The common endpoint is called the **vertex** and the half-lines determined by the rays are called **sides of the angle**. In Figure 3.30 the angle's vertex is point *A* and its sides are $\overrightarrow{AC}$ and $\overrightarrow{AB}$. An angle can be named in several ways. When there is no possibility of confusion, an angle can be named by naming its vertex. The symbol ∠ represents the word "angle." The angle pictured in Figure 3.30 can be designated ∠*A*. Another way to name an angle is to name its vertex and one point on each of its sides. The vertex point is always named between the other two points. In Figure 3.30, the angle can be named ∠*BAC* or ∠*CAB*.

Another way to name an angle is by number. In Figure 3.31, ∠*ABC* can be named ∠1 or ∠*CBA* and ∠*DBC* can be called ∠2 or ∠*CBD*. The third angle in Figure 3.31 is ∠*DBA*. The only other designation for this angle is ∠*ABD*. Note that for the angles in Figure 3.31 we cannot use a one-letter designation such as ∠*B* because it would not be clear which angle we intended.

An angle is determined by three noncollinear points. In fact, three noncollinear points determine several angles. In Figure 3.32, points *A*, *B*, and *C* are noncollinear and they determine the three angles *ABC*, *BCA*, and *CAB*, among others.

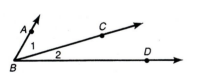

FIGURE 3.31
Naming an angle: ∠2 or ∠*DBC*;
∠1 or ∠*DBC*

FIGURE 3.32
Three angles determined by three noncollinear points: ∠*BAC*, ∠*BCA*, and ∠*CBA*

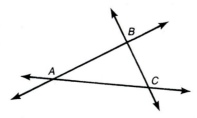

Length or distance may be measured using many different units—feet, inches, miles, meters, and so on. Angles are generally measured in degrees or radians. In this course we only use degree measure. If an angle is drawn so that its vertex is at the center of a circle and the sides of the angle intersect a part of the circle that is $\frac{1}{360}$ as long as the entire length (or distance around the circle), then that angle has a measure of **one degree**. Figure 3.33 shows an angle whose measure is one degree.

FIGURE 3.33
An angle whose measure is one degree

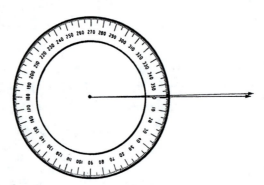

If one of the degree subdivisions is divided into 60 equal parts and an angle is drawn as before, then the measure of this angle is **1 minute**. Similarly, if a subdivision used to measure one minute is subdivided into 60 equal parts and an angle is formed as before, then the measure of this angle is **1 second**.

THE MEASURES OF AN ANGLE

1. One degree equals 60 minutes.
2. One minute equals 60 seconds.
3. One degree equals 3600 seconds.

Degrees are represented by a ° symbol following a number, minutes are represented by a ′ symbol following a number, and seconds are represented by a ″ symbol following the number. Thus, 30° 15′ 24″ is read "30 degrees, 15 minutes, and 24 seconds."

We use an instrument called a protractor to measure angles. Protractors used by most students are marked off only in degrees—minutes and seconds are not shown.

Now let's see how to add and subtract angle measures.

EXAMPLE 2   Add:

$$27° \quad 35′ \quad 53″$$
$$+15° \quad 50′ \quad 17″$$

Solution   To add the measures of two angles, add the seconds first.

$$27° \quad 35′ \quad 53″$$
$$\underline{+15° \quad 50′ \quad 17″}$$
$$70″$$

since $70'' = 1' 10''$, carry one minute and write the $10''$ in the seconds column. Then add the minutes.

$$
\begin{array}{rrr}
& 1' & \\
27° & 35' & 53'' \\
+15° & 50' & 17'' \\
\hline
& 86' & 10''
\end{array}
$$

Since $86' = 1° 26'$, carry $1°$ and write the $26'$ in the minutes column. Then add the degrees.

$$
\begin{array}{rrr}
1° & & \\
27° & 35' & 53'' \\
+15° & 50' & 17'' \\
\hline
43° & 26' & 10''
\end{array}
$$

EXAMPLE 3 Subtract:

$$
\begin{array}{rrr}
49° & 5' & 8'' \\
-36° & 15' & 14'' \\
\hline
\end{array}
$$

Solution To subtract the measures of two angles you may have to borrow. If you do, remember that $1°$ is $60'$ and $1'$ is $60''$.

$$
\begin{array}{rrr}
48° & {}^{6}4' & \\
\cancel{49°} & \cancel{5'} & {}^{6}8'' \\
-36° & 15' & 14'' \\
\hline
12° & 49' & 54''
\end{array}
$$

EXAMPLE 4 Subtract:

$$
\begin{array}{rrr}
90° & & \\
-32° & 15' & 17'' \\
\hline
\end{array}
$$

Solution

$$
\begin{array}{rrr}
89° & 59' & \\
\cancel{90°} & \cancel{60'} & 60'' \\
-32° & 15' & 17'' \\
\hline
57° & 44' & 43''
\end{array}
$$

The measure of an angle is distinguished from the set of points making up the angle by placing an "m" in front of the angle symbol. An angle's actual measure is often placed inside the angle. See Figure 3.34. Note that an angle's measure is not used to name the angle. In Figure 3.34, we write m $\angle A = 35°$; we would never write $\angle 35°$, $\angle A = 35°$, or $A = 35°$. You can tell whether the number placed inside the angle is its measure or its name because its measure in degrees includes a degree symbol.

Be sure you understand the distinction between an angle and its measure. For example, $\angle ABC$ is the set of points in the union of $\overrightarrow{BA}$ and $\overrightarrow{BC}$, whereas m $\angle ABC$ is the measure of the angle and is given in degrees, minutes, and seconds. Thus $\angle ABC$ and m $\angle ABC$ cannot be equal because a set of points cannot be equal to a number.

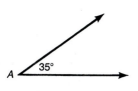

FIGURE 3.34
An angle whose measure is 35°:
m $\angle A = 35°$

2 ▶ ANGLE CLASSIFICATIONS AND RELATIONSHIPS

Two angles are congruent if and only if they have the same measure. If m ∠P = m ∠Q, then ∠P ≅ ∠Q and if ∠P ≅ ∠Q, then m ∠P = m ∠Q. In Figure 3.35, ∠ABC ≅ ∠PQR because they have equal measures.

FIGURE 3.35
∠ABC and ∠PQR are congruent
angles

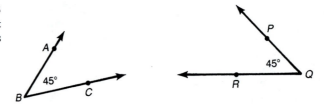

RULES OF SEPARATION FOR AN ANGLE IN A PLANE

An angle divides a plane into three disjoint sets (Figure 3.36):

1. The set of points on the angle
2. The set of points inside the angle, called the angle's **interior**
3. The set of points outside the angle, called the angle's **exterior**

FIGURE 3.36
An angle separates a plane into three
distinct sets

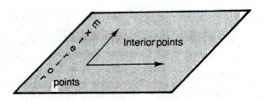

> ### THE ANGLE ADDITION PROPERTY
>
> If a half-line lies in the interior of an angle and its endpoint is the vertex of the angle, then we say the ray it determines divides the angle into two angles, the sum of whose measures is equal to the measure of the given angle.

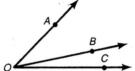

FIGURE 3.37
Angle addition property:
m ∠AOB + m ∠BOC = m ∠AOC

In Figure 3.37, $\overset{\circ}{\overrightarrow{OB}}$ lies in the interior of ∠AOC and it shares its endpoint with the vertex of ∠AOC. Thus $\overrightarrow{OB}$ divides ∠AOC into two angles, the sum of whose measures is equal to the measure of ∠AOC. Specifically,

$$m \angle AOB + m \angle BOC = m \angle AOC$$

Angles are classified in two ways—by their measure and by their relative position. If the measure of an angle is greater than 0° but less than 90°, it is called an **acute angle**. If its measure is greater than 90° but less than 180°, it is called an **obtuse angle**. An angle whose measure is 90° is a **right angle**. Right angles are marked in a figure by placing a square in their interior, as shown in Figure 3.38. An angle whose measure is 180° is called a **straight angle**.

FIGURE 3.38
Classification of angles by measure

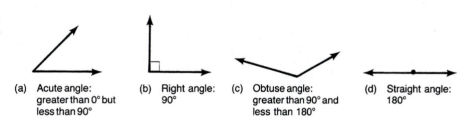

(a) Acute angle: greater than 0° but less than 90°

(b) Right angle: 90°

(c) Obtuse angle: greater than 90° and less than 180°

(d) Straight angle: 180°

If two lines, two line segments, two rays, two half-lines, or any combination of these intersect in such a way that angles measuring 90° are either formed or determined, then they are **perpendicular**. See Figure 3.39.

FIGURE 3.39
Perpendicular lines and subsets

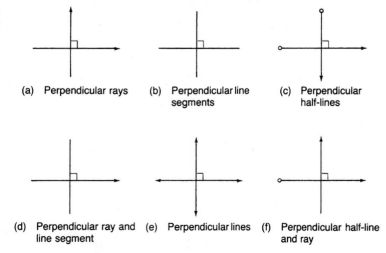

(a) Perpendicular rays

(b) Perpendicular line segments

(c) Perpendicular half-lines

(d) Perpendicular ray and line segment

(e) Perpendicular lines

(f) Perpendicular half-line and ray

QUICK CHECK

Answer each of the following true or false.

1. An angle whose measure is 120° is called an obtuse angle.
2. A right angle must "open" to the right.
3. Rays cannot be perpendicular.
4. An angle whose measure is 180° is called an acute angle.
5. Two angles that have the same measure are congruent.

3 ▶ ADJACENT AND VERTICAL ANGLES

Adjacent angles are angles that have a common vertex and a common side but have no interior points in common. **Vertical angles** share a common vertex and their sides are opposite rays. See Figure 3.40.

FIGURE 3.40
Classification of angles by relative position

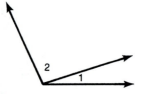
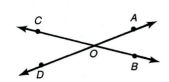

(a) ∠1 and ∠2 are adjacent angles   (b) ∠AOB and ∠COD are vertical angles;
                                           ∠AOC and ∠BOD are vertical angles

If two lines intersect, then two pairs of vertical angles are determined and four pairs of adjacent angles are formed. See Figure 3.41.

FIGURE 3.41
Intersecting lines form vertical and adjacent angles

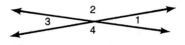

Vertical angles: 1 and 3, 2 and 4
Adjacent angles: 1 and 2, 2 and 3, 3 and 4, 4 and 1

Vertical angles and adjacent angles have two important properties (Figure 3.42):

1. Vertical angles have the same measure.
2. The sum of the measures of adjacent angles formed by intersecting lines is 180°.

ANSWERS
1. True 2. False 3. False 4. False 5. True

FIGURE 3.42

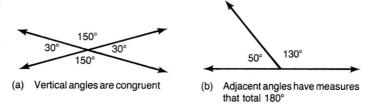

(a) Vertical angles are congruent

(b) Adjacent angles have measures that total 180°

If the sum of the measures of two angles is 90°, the angles are **complementary**. If the sum of the measures of two angles is 180°, they are **supplementary**. See Figure 3.43. Thus, we can conclude that adjacent angles formed by opposite rays are supplementary.

FIGURE 3.43
Complementary and supplementary angles

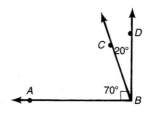

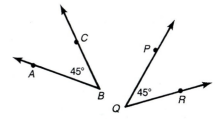

(a) ∠ABC and ∠CBD are complementary because m ∠ABC + m ∠CBD = 90°

(b) ∠ABC and ∠PQR are complementary because m ∠ABC + m ∠PQR = 90°

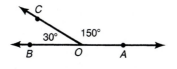

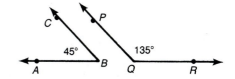

(c) ∠AOC and ∠BOC are supplementary because m ∠AOC + m ∠BOC = 180°

(d) ∠ABC and ∠PQR are supplementary because m ∠ABC + m ∠PQR = 180°

QUICK CHECK

Answer each of the following true or false.

1. Supplementary angles must be acute angles.

2. Adjacent angles are complementary.

3. If m ∠A = 50° and m ∠B = 40°, then the angles are complementary.

4. Vertical angles are congruent.

5. Adjacent angles can each measure 90°.

ANSWERS

1. False   2. False   3. True   4. True   5. True

▶ PARALLEL LINES AND A TRANSVERSAL

If two parallel lines are intersected by a third line (called a **transversal**), then eight angles are determined. These angles are given special names based on their position relative to the transversal and the parallel lines. See Figure 3.44.

FIGURE 3.44

Lines *l* and *n* are parallel; line *t* is the transversal

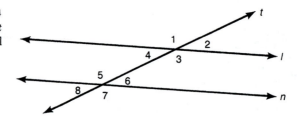

Interior angles: $\angle 3$, $\angle 4$, $\angle 5$, $\angle 6$

Alternate interior angles: $\angle 4$ and $\angle 6$, $\angle 3$ and $\angle 5$

Exterior angles: $\angle 1$, $\angle 2$, $\angle 7$, $\angle 8$

Alternate exterior angles: $\angle 1$ and $\angle 7$, $\angle 2$ and $\angle 8$

Corresponding angles: $\angle 1$ and $\angle 5$, $\angle 4$ and $\angle 8$, $\angle 2$ and $\angle 6$, $\angle 3$ and $\angle 7$

Interior angles on the same side of the transversal: $\angle 4$ and $\angle 5$, $\angle 3$ and $\angle 6$

Exterior angles on the same side of the transversal: $\angle 1$ and $\angle 8$, $\angle 2$ and $\angle 7$

PARALLEL LINES CUT BY A TRANSVERSAL

Two lines intersected by a transversal are parallel if and only if the following congruent and supplementary angles are formed:

Congruent Angles

1. Alternate interior angles
2. Alternate exterior angles
3. Corresponding angles

Supplementary Angles

1. Interior angles on the same side of the transversal
2. Exterior angles on the same side of the transversal

Thus in Figure 3.44

$\angle 4 \cong \angle 6$	$\angle 3 \cong \angle 5$
$\angle 1 \cong \angle 5$	$\angle 4 \cong \angle 8$
$m \angle 4 + m \angle 5 = 180°$	$m \angle 3 + m \angle 6 = 180°$
$\angle 1 \cong \angle 7$	$\angle 2 \cong \angle 8$
$\angle 2 \cong \angle 6$	$\angle 3 \cong \angle 7$
$m \angle 1 + m \angle 8 = 180°$	$m \angle 2 + m \angle 7 = 180°$

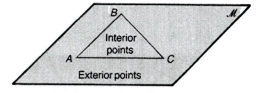

QUICK CHECK

For parallel lines cut by a transversal, answer each of the following true or false.

1. Alternate interior angles are congruent.

2. Interior angles on the same side of the transversal are supplementary.

3. Alternate exterior angles are congruent.

4. Corresponding angles are supplementary.

5. Exterior angles on the same side of the transversal are supplementary.

5 ANGLE BISECTORS AND TRIANGLES

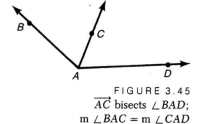

FIGURE 3.45
$\overrightarrow{AC}$ bisects $\angle BAD$;
m $\angle BAC$ = m $\angle CAD$

If the half-line determined by a ray lies in the interior of an angle, the vertex of the angle is the endpoint of the ray, and the ray divides the angle into two angles having the same measure, then the ray **bisects the angle**. In Figure 3.45, $\overrightarrow{AC}$ bisects $\angle BAD$ because m $\angle BAC$ = m $\angle CAD$.

The union of the three line segments determined by three noncollinear points is called a **triangle**. The line segments are called the **sides** of the triangle and the endpoints of the sides are called the **vertices** of the triangle. Figure 3.46 shows a triangle whose vertices are *A*, *B*, and *C* and whose sides are $\overline{AB}$, $\overline{BC}$, and $\overline{AC}$. The symbol Δ is used to represent a triangle. A triangle is named by naming its vertices in any order. The triangle in Figure 3.46 can be named ΔABC, ΔACB, ΔBAC, ΔBCA, ΔCAB, or ΔCBA.

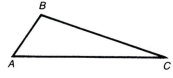

FIGURE 3.46
Triangle *ABC*

RULES OF SEPARATION FOR A TRIANGLE IN A PLANE

A triangle separates a plane into three disjoint sets (Figure 3.47):

1. The set of points on the triangle
2. The set of points inside the triangle, called its interior
3. The set of points outside the triangle, called its exterior

---

FIGURE 3.47
A triangle separates a plane into
three distinct sets

It should be noted that a triangle contains no angles, but its vertices determine three angles whose interior intersects the interior of the triangle. These

FIGURE 3.48
A triangle contains no angles and
determines three angles

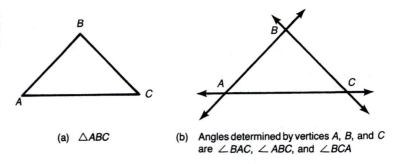

(a)   △*ABC*   (b)   Angles determined by vertices *A*, *B*, and *C*
                            are ∠*BAC*, ∠*ABC*, and ∠*BCA*

angles are called the **angles of the triangle.** When we say that a triangle has three
angles, we are referring to the three angles shown in Figure 3.48.

QUICK CHECK

Answer each of the following true or false.

1. A triangle contains three angles.

2. A triangle determines three angles.

3. An angle bisector is a ray.

4. If a ray bisects an angle, then it divides it into two angles that are
    supplementary.

5. A triangle separates a plane into three disjoint sets.

6 ▶ TRIANGLE CLASSIFICATION

Triangles are classified in two ways—by the relative measures of their sides and
by the measures of their angles.

   If a triangle has no congruent sides, it is called a **scalene triangle.** If it has
at least two congruent sides, it is called an **isosceles triangle.** If all three sides are
congruent, it is called an **equilateral triangle.** Figure 3.49 shows the classification
of triangles with respect to side measure.

FIGURE 3.49
Triangle classification by relative
side length

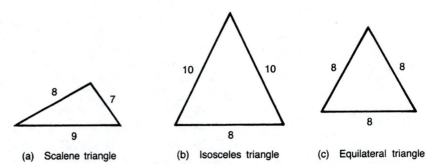

(a) Scalene triangle   (b) Isosceles triangle   (c) Equilateral triangle

If all of the angles of a triangle are acute angles, the triangle is called an **acute triangle**. If a triangle has one right angle, it is called a **right triangle**. If one angle of a triangle is obtuse, the triangle is called an **obtuse triangle**. If all of the angles of a triangle have the same measure, the triangle is called an **equiangular triangle**. Figure 3.50 shows the classification of triangles with respect to angle measure.

FIGURE 3.50
Triangle classification by angle measure

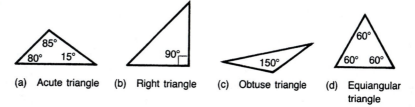

(a) Acute triangle (b) Right triangle (c) Obtuse triangle (d) Equiangular triangle

Triangles have more than one classification, such as an isosceles right triangle or an acute equilateral triangle. Not all combinations are possible, however. For example, as you will see in the next section, you could not have an equiangular right triangle. Figure 3.51 shows several combined classifications. Can you think of others?

FIGURE 3.51
Combined classifications of triangles

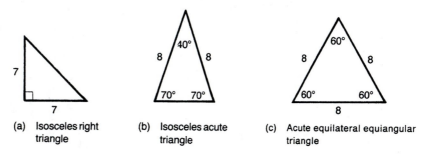

(a) Isosceles right triangle (b) Isosceles acute triangle (c) Acute equilateral equiangular triangle

QUICK CHECK

Answer each of the following true or false.

1. An equilateral triangle has two congruent sides, so it is also an isosceles triangle.

2. A right triangle can have two 90° angles.

3. A right triangle can also be an isosceles triangle.

4. An obtuse triangle has one right angle.

ANSWERS

1. True 2. False 3. True 4. False

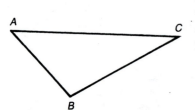

FIGURE 3.52
Adjacent sides and angles of
a triangle

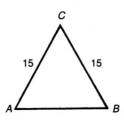

FIGURE 3.53
Opposite sides and angles of
a triangle

FIGURE 3.54
∠A and ∠B are base angles of
isosceles triangle ABC; ∠A ≅ ∠B

7 ▶ ADJACENT SIDES AND ANGLES OF A TRIANGLE

In a triangle pairs of sides that share a common endpoint are called **adjacent sides**, and pairs of angles determined by the triangle are called **adjacent angles**. Figure 3.52 shows the following adjacent angles and sides of a triangle.

Adjacent sides:	$\overline{AB}$ and $\overline{BC}$	Adjacent angles:	∠A and ∠B
	$\overline{AB}$ and $\overline{AC}$		∠B and ∠C
	$\overline{BC}$ and $\overline{AC}$		∠A and ∠C

The side of a triangle that does not have a given vertex as one of its endpoints is called the **side opposite** the angle having the given vertex. Similarly, the angle whose vertex is not an endpoint of a given side is called the **angle opposite** the side. Figure 3.53 shows the following opposite sides and angles of a triangle. The word "opposite" simply means "across from."

$\overline{AB}$ is the side opposite ∠C or ∠C is opposite $\overline{AB}$

$\overline{BC}$ is the side opposite ∠A or ∠A is opposite $\overline{BC}$

$\overline{AC}$ is the side opposite ∠B or ∠B is opposite $\overline{AC}$

In an isosceles triangle, two sides are congruent and the remaining side is called the **base**. The angle opposite the base is called the **vertex angle** and the angles opposite the congruent sides are called the **base angles**. In Figure 3.54, △ABC is isosceles, $\overline{AC}$ and $\overline{BC}$ are the congruent sides, $\overline{AB}$ is the base, C is the vertex angle, and A and B are the base angles. In an isosceles triangle, the base angles are always congruent. Thus, in Figure 3.54 ∠A and ∠B are congruent.

QUICK CHECK

Answer each of the following true or false.

1. In △ABC sides $\overline{AB}$ and $\overline{BC}$ are opposite sides.

2. In △ABC angle C is opposite side $\overline{AB}$.

3. An isosceles triangle has at least two congruent sides and determines at least two congruent angles.

4. An isosceles triangle has two base angles and one vertex angle.

5. A triangle has three pairs of adjacent sides.

---

8 ▶ ALTITUDES AND MEDIANS

In a triangle a line segment drawn from a vertex of the triangle perpendicular to the line determined by the opposite side of the triangle is called the triangle's **altitude** to that side. In Figure 3.55, $\overline{AD}$ is the altitude to $\overline{BC}$ in each triangle. Notice that an altitude does not necessarily lie in the interior of a triangle.

ANSWERS

1. False   2. True   3. True   4. True   5. True

FIGURE 3.55
Altitudes: $\overline{AD}$ is the altitude to $\overline{BC}$ in each triangle

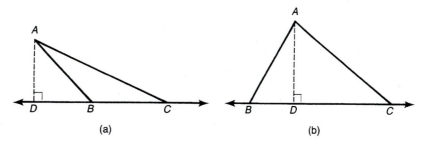

(a)   (b)

In a triangle a line segment drawn from a vertex of the triangle to the midpoint of the opposite side is called the **median** to that side. In Figure 3.56, $\overline{AD}$ is the median to $\overline{BC}$ in each triangle. The median of a triangle must lie in its interior. How many medians does one triangle have?

FIGURE 3.56
Medians: $\overline{AD}$ is the median to $\overline{BC}$ in each triangle

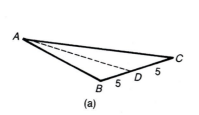   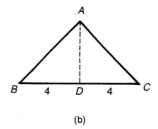

(a)   (b)

QUICK CHECK

Answer each of the following true or false.

1. A median of a triangle bisects an angle of the triangle.

2. In an isosceles triangle the median to the base bisects the side to which it is drawn.

3. Medians must lie in the interior of a triangle.

4. Altitudes of an acute triangle are in the interior of the triangle.

5. A median and an altitude can never be the same line segment.

ANSWERS

1. False 2. True 3. True 4. True 5. False

A

ANSWERS

For Exercises 1–7 refer to the figure shown.

1. Name the angle in three different ways.

2. Name the angle's vertex.

3. Name the rays whose union is the angle.

4. Where is point *P* located?

5. Where is point *Q* located?

6. Where is point *A* located?

7. What is the measure of the angle?

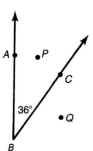

Perform the indicated operations.

8.
$$\begin{array}{rrr} 27° & 15' & 18'' \\ +11° & 17' & 25'' \end{array}$$

9.
$$\begin{array}{rrr} 46° & 15' & 47'' \\ +85° & 23' & 71'' \end{array}$$

10.
$$\begin{array}{rrr} 23° & & 32'' \\ -15° & 14' & 17'' \end{array}$$

11.
$$\begin{array}{rrr} 36° & 45' & \\ -25° & 49' & 23'' \end{array}$$

12.
$$\begin{array}{rrr} 36° & 51' & 11'' \\ -24° & & \end{array}$$

13.
$$\begin{array}{rrr} 90° & & \\ -26° & 43' & 55'' \end{array}$$

14.
$$\begin{array}{rrr} 16° & 53' & 47'' \\ + 8° & 73' & 49'' \end{array}$$

15.
$$\begin{array}{rr} 63° & 47' \\ +193° & 46' \end{array}$$

Answer each of the following true or false.

16. An acute angle measures 90°.

17. An angle whose measure is 180° is called a straight angle.

18. If an angle measures 90°, then it is a right angle.

19. An angle whose measure is 150° is called an obtuse angle.

20. Rays can never be perpendicular.

21. Adjacent angles are congruent.

22. Vertical angles are complementary.

1. _____

2. _____

3. _____

4. _____

5. _____

6. _____

7. _____

8. _____

9. _____

10. _____

11. _____

12. _____

13. _____

14. _____

15. _____

16. _____

17. _____

18. _____

19. _____

20. _____

21. _____

22. _____

23. If one angle measures 10° and another angle measures 80°, then the angles are complementary.

24. If two lines intersect, four pairs of vertical angles are formed.

25. If two lines intersect forming one pair of 90° vertical angles, then the lines are perpendicular.

26. Adjacent angles are always supplementary.

27. The medians of a triangle lie in its interior.

28. The altitudes of a triangle lie in its interior.

29. A triangle determines three angles.

30. A scalene triangle has two congruent sides.

31. An isosceles triangle has at least two congruent sides.

32. An equilateral triangle has exactly two congruent sides.

33. A right triangle can also be equilateral.

For Exercises 34–40 refer to the figure shown.

34. Name the side opposite ∠P.

35. Name the angle opposite $\overline{PQ}$.

36. Where is point A located?

37. Where is point B located?

38. Where is point C located?

39. Name the triangle.

40. Name the vertices.

Draw each of the following.

41. A scalene triangle and its three medians

42. A right triangle and its altitudes

43. An obtuse triangle and its altitudes

44. An acute triangle and its altitudes

45. Supplementary adjacent angles

For Exercises 46–51 refer to the figure shown. Lines l and m and lines p and q are parallel.

46. If m ∠1 = 110°, find the measure of the other angles.

47. Angle 1 is the corresponding angle for which other two angles?

48. Angle 6 is alternate interior with which angle?

49. Angle 6 is alternate exterior to which angle?

50. Are angles 5 and 11 interior angles on the same side of the transversal?

51. Name an angle that is supplementary to angle 5.

23. _____

24. _____

25. _____

26. _____

27. _____

28. _____

29. _____

30. _____

31. _____

32. _____

33. _____

34. _____

35. _____

36. _____

37. _____

38. _____

39. _____

40. _____

41. _____

42. _____

43. _____

44. _____

45. _____

46. _____

47. _____

48. _____

49. _____

50. _____

51. _____

B

52. In the figure shown $\overline{AB} \parallel l$. What is true of the altitudes of all triangles that can be formed with side $\overline{AB}$ and any point C on line l?

52. _____

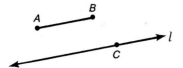

53. Find the measure of $\angle ACB$ in the figure shown.

53. _____

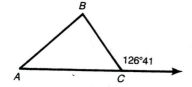

54. If m $\angle 1 = 126° \, 15' \, 43''$, find m $\angle 2$, m $\angle 3$, and m $\angle 4$.

54. _____

55. Three angles measure $26° \, 21' \, 15''$, $21° \, 15' \, 36''$, and $27° \, 21' \, 15''$. What is the sum of their measures?

55. _____

56. Two angles are supplementary. The measure of one angle is three less than twice the other. Find the measure of each angle.

56. _____

57. Two angles are supplementary and adjacent. The measure of one angle is two more than one-half the other. Find the measure of each angle.

57. _____

58. Find m $\angle ABC$ if m $\angle ABD = 70°$, m $\angle EBD = 20°$, m $\angle FBC = 50°$, and m $\angle FBE = $ m $\angle DBC$.

58. _____

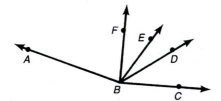

3.3 POLYGONS

OBJECTIVES

▶ **1** Identify a polygon; identify adjacent sides and adjacent angles of a polygon.

▶ **2** Classify and name a polygon and identify its diagonals.

▶ **3** Identify a quadrilateral, identify its adjacent sides and opposite sides, and find the sum of its angles.

▶ **4** Define a parallelogram and state its properties.

▶ **5** Define a rectangle, square, and rhombus and state their properties.

▶ **6** Define a trapezoid and identify its bases, base angles, and height; define an isosceles trapezoid and state the relationship of its base angles.

▶ **7** Find the sum of the angles of any polygon.

▶ **1** POLYGON IDENTIFICATION

Plane figures are sets of points that are coplanar (contained in the same plane). The plane figures in Figure 3.57 are called polygons.

FIGURE 3.57
These are polygons

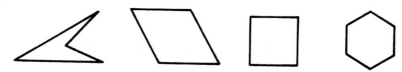

Polygons are plane figures having the following properties:

1. They are formed by the union of line segments called **sides** of the polygon with common endpoints called **vertices** of the polygon.
2. Each endpoint is common to exactly two segments.
3. The sides intersect only at their endpoints.

The plane figures in Figure 3.58 are not polygons.

FIGURE 3.58
These are not polygons

In a polygon, pairs of sides that share common endpoints are called **adjacent sides**. A polygon has or determines as many angles as it has sides or vertices. A polygon's angles are those angles determined by any three consecutive vertices.

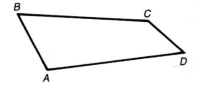

FIGURE 3.59
Adjacent sides and angles of
a polygon

The **adjacent angles** in a polygon are angles of the polygon whose vertices are consecutive vertices of the polygon. The adjacent sides and angles for a polygon having four sides (Figure 3.59) are listed below. This concept is easily extended to include polygons with any number of sides.

Angles A and B, or B and C, or C and D, or D and A are adjacent

Sides $\overline{AB}$ and $\overline{BC}$, or $\overline{BC}$ and $\overline{CD}$, or $\overline{CD}$ and $\overline{DA}$ are adjacent

No polygon other than the triangle has a special symbol. Polygons are named by naming their vertices in any consecutive order. In Figure 3.59, we refer to the polygon as polygon $ABCD$ or $CDAB$, and so on.

QUICK CHECK

Answer each of the following true or false.

1. A polygon determines four angles.
2. A polygon has two pairs of opposite sides.
3. A polygon has two pairs of adjacent sides.
4. The sides and angles of a polygon are coplanar.
5. Adjacent angles of a polygon are supplementary.

▶ **2** CLASSIFYING AND NAMING POLYGONS

A polygon that has all sides congruent is called an **equilateral polygon**; a polygon that has all angles congruent is called an **equiangular polygon**; and a polygon that has all sides and all angles congruent is called a **regular polygon**. See Figure 3.60. Could a polygon be equiangular and not be regular? Yes, a rectangle is an equiangular polygon that need not be regular.

FIGURE 3.60
Equilateral, equiangular, and regular
polygons

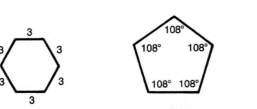

(a) Equilateral polygon   (b) Equiangular polygon   (c) Regular polygon

---

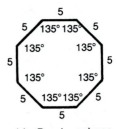

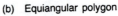

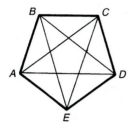

FIGURE 3.61
Diagonals of a polygon: $\overline{AD}$, $\overline{AC}$, $\overline{BE}$, $\overline{BD}$, and $\overline{CE}$ are diagonals of polygon *ABCDE*

Line segments whose endpoints are nonadjacent vertices of a polygon are called **diagonals**. Figure 3.61 shows a five-sided polygon with the diagonals drawn. Do all polygons have diagonals? No, a triangle does not have diagonals because all vertices of a triangle are adjacent.

Polygons are named according to the number of sides they have. Polygons that are not included in the following list have no special names and are called *n*-gons, where *n* is the number of sides. For example, a polygon of 26 sides is called a 26-gon.

NAMES FOR POLYGONS	
Number of Sides	Name of Polygon
3	Triangle
4	Quadrilateral
5	Pentagon
6	Hexagon
7	Heptagon
8	Octagon
9	Nonagon
10	Decagon
12	Dodecagon

In the previous section we discussed the three-sided polygon called the triangle. In a triangle, the sum of the measures of the angles is 180°. See Figure 3.62.

FIGURE 3.62
The sum of the measures of the angles of a triangle is 180°;
m $\angle 1$ + m $\angle 2$ + m $\angle 3$ = 180°

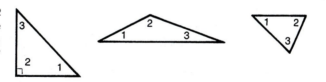

QUICK CHECK

Answer each of the following true or false.

1. A pentagon has five sides.

2. A hexagon has 10 diagonals.

3. A triangle is a polygon.

4. An isosceles triangle is a regular polygon.

5. An equilateral triangle is a regular polygon.

ANSWERS

1. True   2. False   3. True   4. False   5. True

▶ 3   QUADRILATERALS

A **quadrilateral** is a polygon with four sides. There are five classifications of quadrilaterals that we will study in detail—the parallelogram, rectangle, square, rhombus, and trapezoid.

Pairs of nonadjacent sides of a quadrilateral are called **opposite sides**. Pairs of nonadjacent angles of a quadrilateral are called **opposite angles**. Thus a quadrilateral has two pairs of opposite sides and two pairs of opposite angles. See Figure 3.63.

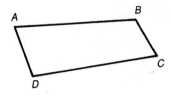

FIGURE 3.63
Opposite sides and angles of a quadrilateral

Pairs of opposite sides: $\overline{AB}$ and $\overline{CD}$, $\overline{AD}$ and $\overline{BC}$

Pairs of opposite angles: $\angle A$ and $\angle C$, $\angle D$ and $\angle B$

A quadrilateral is the only polygon for which opposite sides and angles are defined.

The sum of the measures of the angles of a quadrilateral is 360°. In Figure 3.64,

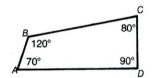

FIGURE 3.64
The sum of the measures of the angles of a quadrilateral is 360°

$$m \angle A + m \angle B + m \angle C + m \angle D = 360°$$

QUICK CHECK

Answer each of the following true or false.

1. The sum of the measures of the angles of a quadrilateral is 360°.

2. A quadrilateral has two pairs of opposite sides.

3. A quadrilateral has two pairs of adjacent sides.

4. Opposite sides of a quadrilateral are congruent.

5. Adjacent angles of a quadrilateral are supplementary.

▶ 4   PARALLELOGRAMS

A **parallelogram** is a quadrilateral in which each pair of nonadjacent (opposite) sides are parallel.

PROPERTIES OF A PARALLELOGRAM
1. The opposite sides are parallel.
2. The opposite sides are congruent.
3. The adjacent angles are supplementary.
4. The opposite angles are congruent.
5. The diagonals bisect each other.

EXAMPLE 5   In the figure shown quadrilateral *ABCD* is a parallelogram. Based on the properties of a parallelogram, list 12 relationships for quadrilateral *ABCD*'s sides and angles.

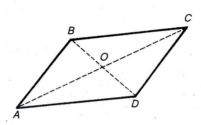

Solution   According to the list of properties for parallelograms, the following relationships are true for quadrilateral *ABCD*.

Since the opposite sides of a parallelogram are parallel,

$$\overline{AD} \parallel \overline{BC} \quad \text{and} \quad \overline{AB} \parallel \overline{CD}$$

Since the opposite sides of a parallelogram are congruent,

$$\overline{AD} \cong \overline{BC} \quad \text{and} \quad \overline{AB} \cong \overline{CD}$$

Since adjacent angles of a parallelogram are supplementary,

$$\text{m} \angle BAD + \text{m} \angle ABC = 180°$$
$$\text{m} \angle ABC + \text{m} \angle BCD = 180°$$
$$\text{m} \angle BCD + \text{m} \angle CDA = 180°$$
$$\text{m} \angle CDA + \text{m} \angle DAB = 180°$$

Since opposite angles of a parallelogram are congruent,

$$\angle BAD \cong \angle BCD \quad \text{and} \quad \angle ABC \cong \angle CDA$$

Since the diagonals bisect each other,

$$\overline{AO} \cong \overline{OC} \quad \text{and} \quad \overline{BO} \cong \overline{OD} \qquad \blacktriangleleft$$

QUICK CHECK

Answer each of the following true or false.

1. A parallelogram is a quadrilateral.

2. A parallelogram can have five sides.

3. The diagonals of a parallelogram are perpendicular.

4. Opposite angles of a parallelogram are congruent as well as supplementary.

5. The diagonals of a parallelogram bisect each other.

ANSWERS

1. True 2. False 3. False 4. False 5. True

THE RECTANGLE, SQUARE, AND RHOMBUS

A **rectangle** is a parallelogram that has four right angles. Thus a rectangle is an **equiangular quadrilateral** and has all of the properties of a parallelogram. See Figure 3.65.

FIGURE 3.65

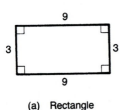

(a)   Rectangle

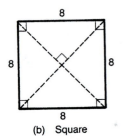

(b)   Square

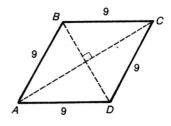

FIGURE 3.66
Rhombus *ABCD*

A **square** is a rectangle with all four sides congruent. Thus a square is a **regular quadrilateral**. The diagonals of a square bisect each other because it is a parallelogram, but the square's diagonals have two additional properties—they are perpendicular and congruent. See Figure 3.65.

A **rhombus** is an equilateral parallelogram. The angles of a rhombus have no restrictions placed on them. The rhombus may or may not be equiangular and it has all of the properties of a parallelogram. In addition, a rhombus has four congruent sides and its diagonals are perpendicular. See Figure 3.66.

QUICK CHECK

Answer each of the following true or false.

1.   A square is a rhombus.   2.   A square is a rectangle.

3.   A rhombus is a rectangle.   4.   A rhombus is a square.

5.   The diagonals of a square are perpendicular.

 TRAPEZOIDS

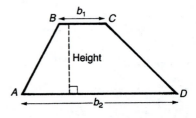

FIGURE 3.67
Trapezoid *ABCD*

A **trapezoid** is a quadrilateral with only one pair of parallel sides. Its parallel sides are called **bases** and are often referred to as b_1 and b_2 (read "base number 1" and "base number 2"). The perpendicular distance between the trapezoid's parallel sides is called the **height**. In general, a pair of **base angles** of a trapezoid is a pair of angles whose vertices are endpoints of one of the bases. Thus a trapezoid has two pairs of base angles. In Figure 3.67, $\angle A$ and $\angle D$ is one pair of base angles and $\angle B$ and $\angle C$ is the second pair of base angles.

If the nonparallel sides of a trapezoid are congruent, it is called an **isosceles trapezoid**. Adjacent angles of an isosceles trapezoid whose intersection is one of

ANSWERS

1. True 2. True 3. False 4. False 5. True

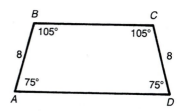

FIGURE 3.68
Angles *A* and *D* or angles *B* and *C*
are base angles of isosceles trapezoid
ABCD

the parallel sides are called base angles of the isosceles trapezoid and they are congruent. An isosceles trapezoid has two pairs of base angles. See Figure 3.68.

QUICK CHECK

Answer each of the following true or false.

1. All trapezoids have two congruent base angles.

2. An isosceles trapezoid has two pairs of congruent base angles.

3. A trapezoid is a quadrilateral.

4. A trapezoid has at least two pairs of parallel sides.

5. The base angles of a trapezoid are supplementary.

7 ▶ THE SUM OF THE ANGLES OF A POLYGON

The sum of the angles determined by a polygon can be determined by dividing the polygon into triangles and using the fact that the sum of the angles of a triangle is 180°. Example 6 explains how the sum of the measures of the angles of a hexagon are obtained. This same technique can be used to find the sum of the measures of the angles of any polygon. If a polygon is equiangular, the measure of each of its angles can be found by dividing the sum of the measures of the angles by the number of angles.

EXAMPLE 6   Find the sum of the measures of the angles of a hexagon, and find the measure of each angle of a regular hexagon.

Solution   To find the sum of the angles of a hexagon, draw triangles having a common vertex at one vertex of the polygon and having sides that are either diagonals or sides of the polygon (refer to the drawing).

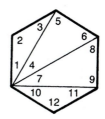

Notice that the sum of the measures of the 12 indicated angles is equal to the sum of the measures of the angles of the hexagon.

$$m \angle 1 + m \angle 2 + m \angle 3 = 180°$$
$$m \angle 4 + m \angle 5 + m \angle 6 = 180°$$
$$m \angle 7 + m \angle 8 + m \angle 9 = 180°$$
$$m \angle 10 + m \angle 11 + m \angle 12 = 180°$$

Therefore, the sum of the measures of the angles of the hexagon is given by

$$4 \cdot 180° = 720°$$

If the hexagon is equiangular, then each angle measures

$$\frac{720°}{6} = 120°$$

◀

ANSWERS

1. False   2. True   3. True   4. False   5. False

EXERCISE 3.3

A

ANSWERS

For Exercises 1–22 match each of the following to the appropriate drawing. A description may match one figure, more than one figure, or no figure.

1. Equilateral triangle
2. Equiangular triangle
3. Parallelogram
4. Trapezoid
5. Scalene triangle
6. Isosceles triangle
7. Rectangle
8. Square
9. Right triangle
10. Obtuse triangle
11. Right angle
12. Obtuse angle
13. Acute triangle
14. Acute angle
15. Isosceles trapezoid
16. Rhombus
17. Hexagon
18. Regular polygon
19. Octagon
20. Pentagon
21. Decagon
22. Quadrilateral

a.
b.
c.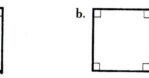
d.
e.
f.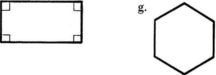
g.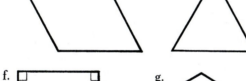
h.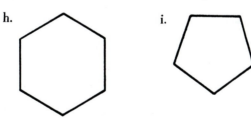
i.
j.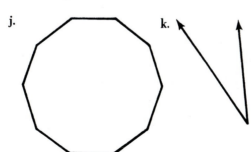
k.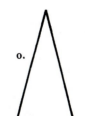
l.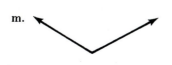
m.
n.
o.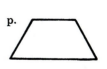
p.

1. _____
2. _____
3. _____
4. _____
5. _____
6. _____
7. _____
8. _____
9. _____
10. _____
11. _____
12. _____
13. _____
14. _____
15. _____
16. _____
17. _____
18. _____
19. _____
20. _____
21. _____
22. _____

23. Which of the following are polygons?

23. _____

a.
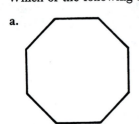

b.

c.

d.
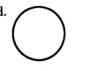

e.

f.

24. What is the sum of the measures of the angles of an octagon?

24. _____

25. What is the measure of each angle of a regular dodecagon?

25. _____

26. Adjacent angles of a parallelogram are supplementary. Why?

26. _____

27. In the figure shown, find the measure of each angle.

27. _____

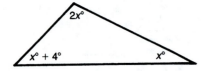

Answer each of the following true or false.

28. Adjacent sides of a polygon do not intersect.

28. _____

29. A pentagon has six sides.

29. _____

30. A star is a pentagon.

30. _____

31. The sum of the measures of the angles of a regular dodecagon is 1080°.

31. _____

32. Equilateral means having sides of equal measure.

32. _____

33. There is no two-sided polygon.

33. _____

34. Opposite sides of a parallelogram are congruent.

34. _____

35. A square is a rhombus.

35. _____

36. A square is a rectangle.

36. _____

37. Base angles of an isosceles trapezoid are congruent.

37. _____

38. A trapezoid has two pairs of base angles.

38. _____

For Exercises 39–44 find the value of the unknown.

39.

39. _____

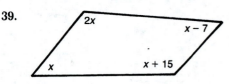

40.

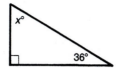

40. _____

41.

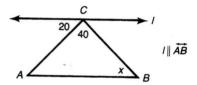

41. _____

42.

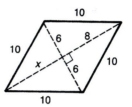

$l \parallel \overleftrightarrow{AB}$

42. _____

43.

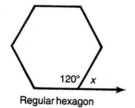

Regular hexagon

43. _____

44.

Regular decagon

44. _____

B

45. In the figure shown line l is parallel to $\overleftrightarrow{AB}$. Show that the sum of the measures of the angles of $\triangle ABC$ must be 180°. Hint: Start with m $\angle 1$ + m $\angle 2$ + m $\angle 3$ = 180° and find angles whose measures are equal to the measures of angles 1 and 3. Then substitute these values into the equation.

45. _____

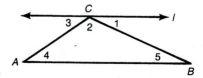

3.4 CIRCLES

OBJECTIVES

▶**1** Define a circle; identify a radius and a diameter; define congruent circles.

▶**2** Identify tangent and secant lines; define tangent and concentric circles.

▶**3** Differentiate between inscribed and circumscribed geometrical figures; identify an angle inscribed in a circle and a central angle of a circle.

▶**4** Define and name an arc and state how an arc is measured; differentiate between a major and a minor arc; define a semicircle.

▶**5** Find the measure of an inscribed angle.

▶**6** Find the measure of angles that intersect a circle.

▶**1** CIRCLES, DIAMETERS, CHORDS, AND RADII

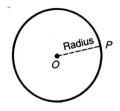

FIGURE 3.69
O is the center of circle O;
$\overline{OP}$ is the radius

A **circle** is the set of all points in a plane that are a given distance from a given point in the plane. The given distance is the **radius** of the circle and the given point is the **center** of the circle. The line segment determined by the center of a circle and any point on the circle is also called its **radius** (plural is radii). The symbol ⊙ is used with the center point to name a circle. The circle in Figure 3.69 is referred to as ⊙O, which is read "circle O."

A **chord** is a line segment whose endpoints are any two points on the circle. A chord that contains the center of the circle is called a **diameter** of the circle. The length of a chord containing the diameter is also called the diameter of the circle. In Figure 3.70, $\overline{AB}$ is a chord and $\overline{PQ}$ is a diameter. In a circle the length of the diameter is twice the length of the radius.

Diameter length = 2 · Radius length

Two circles are congruent if and only if they have equal radii or equal diameters.

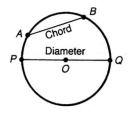

FIGURE 3.70
Chords

QUICK CHECK

Answer each of the following true or false.

1. A diameter is a chord.

2. If the diameter of a circle measures 8 inches, its radius measures 4 inches.

3. A radius is a chord.

4. A radius is a line segment.

5. The points of a circle are coplanar.

6. Circles with congruent radii are congruent.

ANSWERS

1. True   2. True   3. False   4. True   5. True   6. True

2 ▶ TANGENT AND SECANT LINES; TANGENT
AND CONCENTRIC CIRCLES

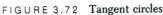

Tangent line

Radius

O

Secant line

FIGURE 3.71
Tangent and secant lines; the radius
is perpendicular to the tangent line

If a line and a circle are coplanar and if the line intersects the circle in exactly
one point, then the line and the circle are tangent at their point of intersection
and the line is called a **tangent line**. A radius drawn to the point of tangency is
perpendicular to the tangent line. A line that intersects a circle in two points is
called a **secant line**. See Figure 3.71.

Circles that are coplanar and intersect in exactly one point, are called **tangent
circles**. Figure 3.72 shows three circles that are tangent at point *P*.

Coplanar circles that have the same center are called **concentric circles**. See
Figure 3.73.

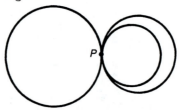

P

O

FIGURE 3.72   Tangent circles   FIGURE 3.73   Concentric circles

---

QUICK CHECK

Answer each of the following true or false.

1. Concentric circles must be coplanar.

2. A radius of a circle is perpendicular to a tangent line at the point of
   tangency.

3. Concentric circles intersect in exactly one point.

4. Tangent circles intersect in exactly one point and are coplanar.

5. Tangent circles must be coplanar.

---

3 ▶ INSCRIBED AND CIRCUMSCRIBED
GEOMETRICAL FIGURES

If all of the vertices of a polygon lie on a circle, then the polygon is **inscribed** in
the circle and the circle is **circumscribed** about the polygon. In Figure 3.74, polygon *ABCDE* is inscribed in ⊙*O* and ⊙*O* is circumscribed about polygon *ABCDE*.

FIGURE 3.74
A polygon inscribed in a circle and
a circle circumscribed about a polygon

B

C

A

O

D

E

A **circle inscribed** in a polygon is tangent to each of the sides of the polygon.
In Figure 3.75, ⊙*O* is inscribed in polygon *ABCD* and polygon *ABCD* is circumscribed about ⊙*O*.

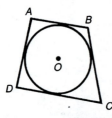

A

B

O

D

C

FIGURE 3.75
A circle inscribed in a polygon

ANSWERS

1. True   2. True   3. False   4. True   5. True

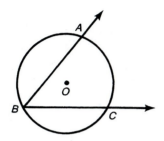

FIGURE 3.76
Angle *ABC* is inscribed in the circle

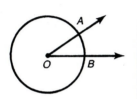

FIGURE 3.77
A central angle

An **inscribed angle** is an angle whose vertex is on a circle and whose sides intersect the circle. In Figure 3.76, $\angle ABC$ is inscribed in $\odot O$.

An angle coplanar with a circle whose vertex is the center of a circle is called a **central angle** of the circle. In Figure 3.77, $\angle AOB$ is a central angle of $\odot O$.

QUICK CHECK

Answer each of the following true or false.

1. An angle inscribed in a circle intersects the circle in exactly three points.

2. A central angle of a circle intersects the circle in at least three points.

3. A circle cannot be circumscribed about a triangle.

4. Inscribed means "on the inside of."

5. Circumscribed means "on the outside of."

▶ 4 ARCS AND THEIR MEASURE

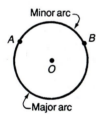

FIGURE 3.78
A and *B* divide circle *O* into major and minor arcs

Any two points on a circle divide the circle into two sets called **arcs**. Thus an arc is a subset of a circle. If two points on a circle are not endpoints of a diameter, then they divide the circle into a **major arc** and a **minor arc**. In Figure 3.78 points *A* and *B* divide $\odot O$ into a major and a minor arc.

If two points on a circle are endpoints of a diameter, then the arcs that they determine are called **semicircles**. In Figure 3.79 line *AB* is a diameter and points *A* and *B* determine two semicircles.

An arc is named by using the symbol $\frown$ and naming three points on the arc. Thus, $\overset{\frown}{ABC}$ is read "arc *ABC*." In Figure 3.80, the minor arc determined by points *A* and *C* is $\overset{\frown}{ABC}$ and the major arc is $\overset{\frown}{ADC}$.

The degree measure of an arc is determined by the measure of its central angle. To designate the measure of an arc, precede the arc symbol by an m. Thus, in Figure 3.81, $\overset{\frown}{mAPB} = 30°$. Angle *AOB* is the central angle whose sides contain the endpoints of the arc it measures.

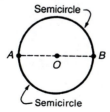

FIGURE 3.79
Arcs whose endpoints lie on a diameter are semicircles

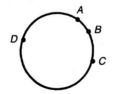

FIGURE 3.80
$\overset{\frown}{ABC}$ names arc *ABC*

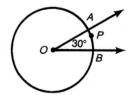

FIGURE 3.81
An arc is measured by its central angle: $\overset{\frown}{mAPB} = 30°$

ANSWERS

1. True 2. False 3. False 4. True 5. True

QUICK CHECK

Answer each of the following true or false.

1. An arc contains exactly three points.   2. $^m\widehat{PQR}$ is read "arc *PQR*."

3. $^m\widehat{PQR} = \widehat{PQR}$.   4. A semicircle is an arc.

5. An arc may be measured in degrees.

6. An arc has the measure of the central angle whose sides contain the endpoints of the arc.

7. A semicircle measures 180°.

5 ▶ MEASURE OF AN INSCRIBED ANGLE

The measure of an inscribed angle is equal to one-half the measure of the arc it cuts off or intercepts. In Figure 3.82, m $\angle APB = (\frac{1}{2})^m\widehat{ARB}$.

If an inscribed angle intercepts a semicircle, then the angle is a right angle because the measure of a semicircle is 180°. The measure of an angle inscribed in a semicircle is equal to one-half the measure of the semicircle and therefore measures 90°. In Figure 3.83, $\angle ACB$ intercepts semicircle *ADB*, so it is a right angle.

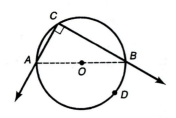

FIGURE 3.82
The measure of an inscribed angle:
m $\angle APB = (\frac{1}{2})^m\widehat{ARB}$

FIGURE 3.83
An angle inscribed in a semicircle
is a right angle

QUICK CHECK

Answer each of the following true or false.

1. Any angle inscribed in a semicircle measures 90°.

2. A central angle and an inscribed angle intercepting the same arc have the same measure.

For Questions 3–5 refer to the figure shown.

3. m $\angle ABC = 80°$

4. m $\angle ABC = 40°$

5. m $\angle ABC = 20°$

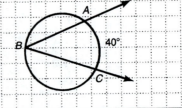

6 ▶ MEASURES OF ANGLES THAT INTERSECT A CIRCLE

If the sides of an angle intersect a circle in four points, then the measure of the angle is one-half the difference between the measures of the arcs it intercepts. In Figure 3.84, $\angle A$ intersects $\odot O$ at points B, D, E, and G, so m $\angle A = (\frac{1}{2})(\overset{\frown}{^mEFG} - \overset{\frown}{^mBCD})$.

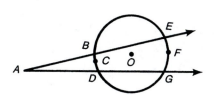

FIGURE 3.84
Measure of an angle that intersects a circle in four points:
m $\angle A = (\frac{1}{2})(\overset{\frown}{^mEFG} - \overset{\frown}{^mBCD})$

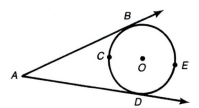

FIGURE 3.85
Measure of an angle whose sides are tangent to a circle:
m $\angle A = (\frac{1}{2})(\overset{\frown}{^mBED} - \overset{\frown}{^mBCD})$

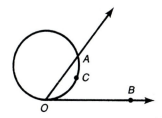

FIGURE 3.86
Measure of an angle that determines a secant and a tangent whose vertex is on the circle: m $\angle AOB = (\frac{1}{2})(\overset{\frown}{^mACO})$

If a circle is tangent to two sides of an angle, then the circle is inscribed in the angle. The measure of the angle is one-half the difference of the measures of the arcs it intercepts. In Figure 3.85, $\odot O$ is inscribed in $\angle A$ and m $\angle A = (\frac{1}{2})(\overset{\frown}{^mBED} - \overset{\frown}{^mBCD})$.

If the vertex of an angle lies on a circle, one of its sides intersects the circle in one point, and the other side is a subset of a tangent line to the circle, then the measure of the angle is one-half the measure of the arc it intercepts. In Figure 3.86 m $\angle AOB = (\frac{1}{2})(\overset{\frown}{^mACO})$. Note that $\overrightarrow{OA}$ is a subset of secant $\overleftrightarrow{OA}$ and that $\overrightarrow{OB}$ is a subset of tangent $\overleftrightarrow{OB}$.

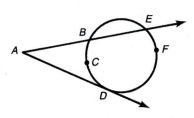

FIGURE 3.87
Measure of an angle that determines a secant and a tangent whose vertex is not on the circle:
m $\angle A = (\frac{1}{2})(\overset{\frown}{^mEFD} - \overset{\frown}{^mBCD})$

If the vertex of an angle is on the exterior of a circle, one side of the angle is tangent to a circle, and the other side intersects it at two points, then the measure of the angle is one-half the difference of the two arcs it intercepts. In Figure 3.87 m $\angle A = (\frac{1}{2})(\overset{\frown}{^mEFD} - \overset{\frown}{^mBCD})$. Note that $\overrightarrow{AE}$ is a subset of secant $\overleftrightarrow{AE}$ and $\overrightarrow{AD}$ is a subset of tangent $\overleftrightarrow{AD}$.

EXAMPLE 7

Find the measure of the unknown in each of the following figures.

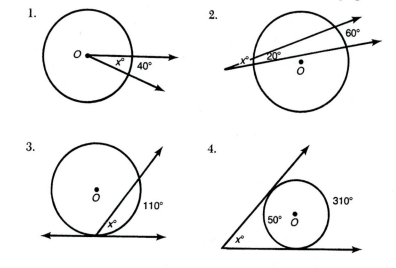

5. 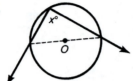   **6.**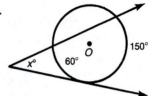

Solution   **1.** $x = 40°$ **2.** $x = (\frac{1}{2})(60° - 20°)$ **3.** $x = (\frac{1}{2})(110°)$
                                          $= (\frac{1}{2})(40°)$ $= 55°$
                                          $= 20°$

4. $x = (\frac{1}{2})(310° - 50°)$ **5.** $x = 90°$ **6.** $x = (\frac{1}{2})(150° - 60°)$
   $= (\frac{1}{2})(260°)$ $= (\frac{1}{2})(90°)$
   $= 130°$ $= 45°$

A

ANSWERS

Refer to the figure shown and name each of the following with respect to circle O.

1. $\overline{OD}$

2. $\angle DOA$

3. $\angle CBA$

4. l

5. $\overrightarrow{EF}$

6. $\overline{EF}$

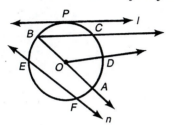

1. _____

2. _____

3. _____

4. _____

5. _____

6. _____

Draw each of the following.

7. A circle inscribed in a pentagon

8. An equilateral triangle inscribed in a circle

9. Five concentric circles

10. Five tangent circles

11. A circle circumscribed about an isosceles triangle

7. _____

8. _____

9. _____

10. _____

11. _____

Answer each of the following true or false.

12. A secant is a line that intersects a circle in exactly one point.

13. A tangent is a line that intersects a circle in exactly one point.

14. Central angles are inscribed in a circle.

15. Two points on a circle divide the circle into a major and a minor arc.

16. A chord is a line segment that intersects a circle in two points.

17. A chord is a diameter.

18. A diameter is a chord.

19. If a diameter of a circle measures 4 inches, then its radius measures 8 inches.

20. An arc is measured by its central angle.

21. If two circles are congruent, then their diameters have the same length.

12. _____

13. _____

14. _____

15. _____

16. _____

17. _____

18. _____

19. _____

20. _____

21. _____

Find the value of the unknown for each of the following.

22.

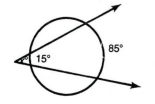

22. _____

23.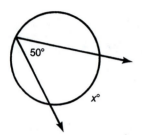

24.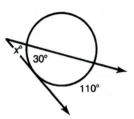

25.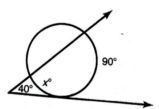

26.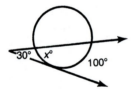

27.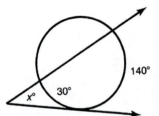

28.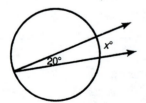

29.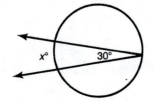

30.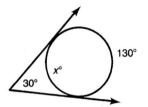

30. _____

31.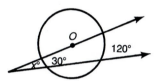

31. _____

32.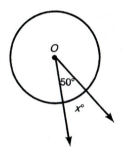

32. _____

33.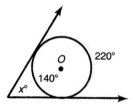

33. _____

B

In the figure shown O is the center of the circle and m $\angle D = 40°$. *Find each of the following.*

34. m $\angle AOC$

35. m $\angle ABC$

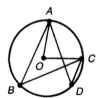

34. _____

35. _____

In the figure shown ABCD is a quadrilateral inscribed in circle O and m $\angle BAD = 100°$, m$\widehat{ARD} = 20°$, *and* m$\widehat{DSC} = 90°$.

36. m $\angle ABC$

37. m $\angle ADC$

38. m $\angle BCD$

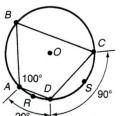

36. _____

37. _____

38. _____

3.5 SPACE FIGURES: PRISMS, PYRAMIDS, CONES, CYLINDERS, AND SPHERES

OBJECTIVES

▶**1** Define a space figure; identify a prism, its bases, edges, and lateral faces; classify a prism by identification of its base type; identify a right prism and an oblique prism.

▶**2** Identify a pyramid, its base, lateral faces, edges, and vertex; classify a pyramid by its base type; distinguish between a right and an oblique pyramid.

▶**3** Identify a cone; distinguish between a right and an oblique cone.

▶**4** Identify a cylinder; distinguish between a right and an oblique cylinder.

▶**5** Define a sphere and identify its radii, diameters, and chords; define a great circle.

▶**6** Identify heights of prisms, pyramids, cones, and cylinders; identify slant heights of cones and pyramids.

▶**1** SPACE FIGURES AND PRISMS

Space figures are sets of points that are not coplanar; that is, space figures are three dimensional. In this section we discuss five kinds of space figures—prisms, pyramids, cones, cylinders, and spheres.

Congruent polygons have the same number of sides and angles and their corresponding sides and angles are congruent. A **prism** is a geometrical space figure that has two congruent polygonal **bases** that lie in parallel planes. The corresponding vertices of these polygons are endpoints of line segments that are called **edges** of the prism. See Figure 3.88.

FIGURE 3.88
Prisms

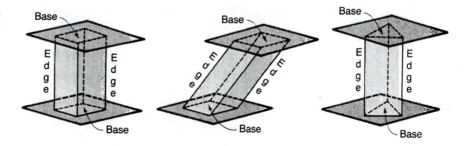

The sides of a prism are either nonrectangular parallelograms or they are rectangles. These sides are called **lateral faces**. If some of the lateral faces are nonrectangular parallelograms, the prism is an **oblique prism**. If the lateral faces are all rectangles, the prism is a **right prism**. See Figure 3.89, page 138.

Prisms are identified by their base type. For example, a prism with hexagonal bases is called a hexagonal prism and a prism with triangular bases is called a triangular prism. Exceptions to this rule are those prisms with bases that are

FIGURE 3.89

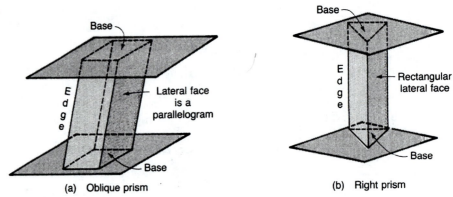

(a)   Oblique prism

(b)   Right prism

parallelograms (or rectangles), in which case the prisms are called **parallelepipeds**. Prisms are classified as follows.

CLASSIFICATION OF PRISMS	
Base Type	Prism Identification
Triangle	Triangular prism
Parallelogram	Parallelepiped
Rectangle	Rectangular parallelepiped
Trapezoid	Trapezoidal prism
Rhombus	Rhombic prism
Quadrilateral	Quadrilateral prism
Pentagon	Pentagonal prism
Hexagon	Hexagonal prism
Heptagon	Heptagonal prism
Octagon	Octagonal prism
Nonagon	Nonagonal prism
Decagon	Decagonal prism
Dodecagon	Dodecagonal prism
n-gon	n-gonal prism

QUICK CHECK

Answer each of the following true or false.

1. A parallelepiped is a prism whose lateral faces are rectangles.

2. A parallelepiped is a prism whose bases are parallelograms or rectangles.

3. A hexagonal prism has 6 lateral faces.

4. The bases of a prism lie in parallel planes.

5. The bases of a prism are congruent polygons.

ANSWERS

1. False   2. True   3. True   4. True   5. True

2 ▶ PYRAMIDS

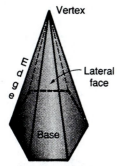

(a) Right hexagonal pyramid

A **pyramid** is a space figure that has one polygonal base and triangular **lateral faces** that share a common vertex. This common vertex is called the **vertex** of the pyramid. See Figure 3.90. If the triangular faces of a pyramid are isosceles, then the pyramid is a **right pyramid**. If the triangular faces are not isosceles, then the pyramid is an **oblique pyramid**. The sides of the lateral faces sharing one endpoint with the vertex are called **edges**. A right pyramid and an oblique pyramid are shown in Figure 3.90.

Pyramids are identified by their base type. Pyramids are classified as follows.

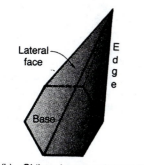

(b) Oblique hexagonal pyramid

FIGURE 3.90

CLASSIFICATION OF PYRAMIDS	
Base Type	Pyramid Identification
Triangle	Triangular pyramid
Parallelogram	Parallelogram pyramid
Rectangle	Rectangular pyramid
Trapezoid	Trapezoidal pyramid
Rhombus	Rhombic pyramid
Quadrilateral	Quadrilateral pyramid
Pentagon	Pentagonal pyramid
Hexagon	Hexagonal pyramid
Heptagon	Heptagonal pyramid
Octagon	Octagonal pyramid
Nonagon	Nonagonal pyramid
Decagon	Decagonal pyramid
Dodecagon	Dodecagonal pyramid
n-gon	n-gonal pyramid

QUICK CHECK

Answer each of the following true or false.

1. A pyramid has two bases.
2. The vertex of a pyramid is a point.
3. A pyramid is classified by its base type.
4. A box is a rectangular pyramid.
5. An octagonal pyramid has eight lateral faces.

ANSWERS

1. False 2. True 3. True 4. False 5. True

▶ 3 CONES

A **circular cone** is the union of all lines determined by the points on a circle and a point not in the plane of the circle. See Figure 3.91.

FIGURE 3.91
Circular cone

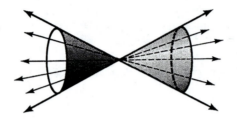

Vertex

Base

FIGURE 3.92
Circular cone

In common practice the subset of the cone represented in Figure 3.92 is referred to as a **circular cone**. It is the union of all line segments determined by the points on a circle and a point not in the plane of the circle. The circle is called the **base** of the cone and the point is called its **vertex**.

If a line segment drawn from the vertex of the cone to the center of its circular base is perpendicular to all of the circle's diameters, the cone is a **right circular cone**; if a line segment drawn from the vertex of the cone to the center of its circular base is not perpendicular to all of the circle's diameters, the cone is an **oblique circular cone**. See Figure 3.93.

FIGURE 3.93

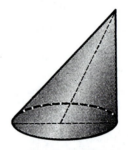

(a)   Right circular cone   (b)   Oblique circular cone

QUICK CHECK

Answer each of the following true or false.

1.   A circular cone has a circular base.

2.   The vertex of a cone is a point.

3.   Oblique cones tilt to one side.

4.   In a right cone a line segment from its vertex to the center of the base is perpendicular to every diameter of the base.

ANSWERS

1. True   2. True   3. True   4. True

▶ CYLINDERS

A **circular cylinder** is the union of all lines that are parallel to the line determined by the centers of two parallel congruent circles, and each line intersects exactly one point of each circle. The circles are called **bases** of the cylinder. See Figure 3.94.

FIGURE 3.94
Circular cylinder

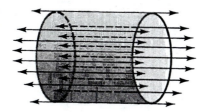

As with a cone, it is common practice to call the subset of the cylinder that is determined by line segments a cylinder. See Figure 3.95.

If the line segment determined by the centers of the circles is perpendicular to their diameters, the cylinder is called a **right circular cylinder**. If the line segment is not perpendicular to the diameters, the cylinder is called an **oblique circular cylinder**. See Figure 3.96.

FIGURE 3.95
Cylinder

FIGURE 3.96

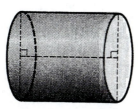

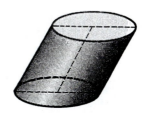

(a) Right circular cylinder   (b) Oblique circular cylinder

QUICK CHECK

Answer each of the following true or false.

1. A circular cylinder has two circular bases.

2. A line joining the centers of the bases of a right circular cylinder is perpendicular to all diameters of its bases.

3. The bases of a circular cylinder lie in parallel planes.

4. A soup can without a top or bottom is an example of a circular cylinder.

ANSWERS

1. True   2. True   3. True   4. True

5 ▶ SPHERES

FIGURE 3.97
Sphere

A **sphere** is the set of all points in space that are a given distance from a given point. The given point is called the **center** of the sphere and the given distance is called the **radius** of the sphere. As with circles, the term radius refers to a line segment determined by the center of the sphere and a point on the sphere as well as to the length of such a line segment. A **chord** of a sphere is any line segment determined by two points on the sphere. A chord that contains the center of the sphere is called the **diameter** of the sphere. The length of a diameter is also called the diameter of the sphere. Figure 3.97 shows a chord, diameter, and radius of a sphere.

If a plane and a sphere intersect in such a way that the plane contains the center of the sphere, then their intersection is called a **great circle**. In Figure 3.98, O is a great circle of the sphere.

FIGURE 3.98
Great circle O

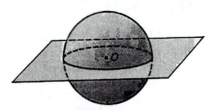

Assuming that the earth is a sphere, the equator would be a great circle. Longitudinal lines are also great circles, but latitude lines are not. See Figure 3.99.

FIGURE 3.99

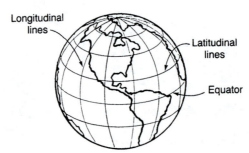

QUICK CHECK

Answer each of the following true or false.

1. The points of a sphere are coplanar.

2. A chord of a sphere is a diameter.

3. The equator is a great circle.

4. All longitudinal lines are great circles.

5. All latitude lines are great circles.

ANSWERS

1. False   2. False   3. True   4. True   5. False

HEIGHTS OF SPACE FIGURES

The **height of a prism or cylinder** is the perpendicular distance between its bases; the **height of a cone or a pyramid** is the perpendicular distance from its vertex to its base. Figure 3.100 pictures four right space figures having equal height.

FIGURE 3.100
Heights

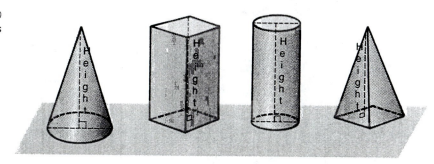

Pyramids with regular bases and cones have **slant heights**. The slant height of a cone is the distance from its vertex to a point on its circular base; the slant height of a pyramid is the length of the altitude of its triangular face. See Figure 3.101.

FIGURE 3.101
Slant heights

 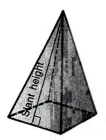

QUICK CHECK

Answer each of the following true or false.

1. The height of a space figure is the length of a particular line segment.
2. The slant height for each face of an oblique pyramid can be different.
3. A cylinder has a slant height.
4. Cones and pyramids have slant heights.
5. The height of a right cylinder is the length of the line segment determined by the centers of its bases.

ANSWERS

1. True 2. True 3. False 4. True 5. True

EXERCISE 3.5

A

ANSWERS

Answer each of the following true or false.

1. A cylinder has a vertex.

2. Latitude lines are great circles.

3. The planes containing the bases of a prism intersect.

4. A triangular prism has three lateral faces and two bases.

5. The lateral faces of prisms are rectangles.

6. Lateral faces of right pyramids are isosceles triangles.

7. Lateral faces of right pyramids are equilateral triangles.

8. A prism whose bases are 13-gons is called a 13-gonal prism.

9. The radius length of a sphere is half its diameter's length.

10. The slant height of a right pyramid is the median of its lateral face.

1. _____

2. _____

3. _____

4. _____

5. _____

6. _____

7. _____

8. _____

9. _____

10. _____

For Exercises 11–20, draw the indicated geometrical figure.

11. A rectangular parallelepiped

12. A pentagonal pyramid

13. A cone

14. A sphere

15. A heptagonal prism

16. A prism with a trapezoidal base

17. A pyramid with an isosceles triangular base

18. A plane intersecting a sphere

19. An oblique cylinder

20. An oblique cone

21. How many lateral faces does a heptagonal prism have?

22. How many lateral faces does a nonagonal pyramid have?

23. How many lateral faces does a triangular pyramid have?

24. How many lateral faces does a decagonal prism have?

25. How many lateral faces does an octagonal pyramid have?

11. _____

12. _____

13. _____

14. _____

15. _____

16. _____

17. _____

18. _____

19. _____

20. _____

21. _____

22. _____

23. _____

24. _____

25. _____

B

Draw each of the following.

26. A plane intersecting a cone. Draw the plane parallel to the base of the cone.

27. A plane intersecting a rectangular pyramid. Draw the plane parallel to the base of the pyramid.

26. _____

27. _____

28. A sphere tangent to a plane.   28. _____

29. Two intersecting planes that are tangent to a sphere.   29. _____

30. A sphere inscribed in a pyramid. (The sphere will be tangent to the faces   30. _____
     of the pyramid.)

Write each of the following in words.

ANSWERS

1. $\triangle ABC$

1. _____

2. $\overset{\frown}{PDQ}$

2. _____

3. $\overrightarrow{AB}$

3. _____

4. BD

4. _____

5. $m\overset{\frown}{RST}$

5. _____

6. $\overline{AZ}$

6. _____

7. $\overleftrightarrow{VT}$

7. _____

8. $\overrightarrow{RK}$

8. _____

9. $\angle A$

9. _____

10. $m \angle PTZ$

10. _____

Fill in the blanks with the correct word or phrase.

11. If two planes intersect, their intersection is a _____

11. _____

12. Noncoplanar lines are called _____

12. _____

13. A _____ is a half-line that contains its endpoint.

13. _____

14. A _____ prism has 12 lateral faces.

14. _____

15. _____ points are points that belong to a line.

15. _____

16. A _____ is the set of all points.

16. _____

17. _____ is determined by two points.

17. _____

18. A _____ is determined by three noncollinear points.

18. _____

19. Two points are _____ as well as _____

19. _____

20. Coplanar lines that do not intersect are _____

20. _____

21. _____ means to divide into two parts having the same measure.

21. _____

22. _____ rays are collinear and have a common endpoint.

22. _____

23. An _____ is the union of two rays with a common endpoint.

23. _____

24. The vertices of a triangle determine _____ _____

24. _____

25. Angles having the _____ are congruent.

25. _____

26. An _____ is an angle whose measure is less than 90° and greater than 0°.

26. _____

27. A right angle measures _____ degrees.

27. _____

28. An angle whose measure is 180° is called a _____

28. _____

29. An angle whose measure is greater than 90° but less than 180° is called an _____ _____.

29. _____

30. Two lines are _____ if they intersect forming a right angle.

30. _____

31. If the sum of the measures of two angles is 90°, then they are _____.

31. _____

32. The sum of the measures of two supplementary angles is _____.

32. _____

33. An _____ triangle has two congruent sides.

33. _____

34. Scalene triangles have _____ congruent sides.

34. _____

35. The measure of each angle of an equilateral triangle is _____.

35. _____

36. A right triangle is a triangle containing _____.

36. _____

37. A polygon with four sides is called _____.

37. _____

38. A polygon with congruent sides and angles is _____.

38. _____

39. The sum of the measures of the angles of a triangle is _____.

39. _____

40. The sum of the measures of the angles of a regular pentagon is _____.

40. _____

41. A _____ is a quadrilateral with two pairs of parallel sides.

41. _____

42. A _____ is a quadrilateral with exactly one pair of parallel sides.

42. _____

43. A _____ is an equilateral parallelogram.

43. _____

44. A _____ is a regular parallelogram.

44. _____

45. Adjacent angles of a parallelogram are _____.

45. _____

46. Opposite angles of a parallelogram are _____.

46. _____

47. The diagonals of a rhombus are _____.

47. _____

48. An _____ is a polygon with eight sides.

48. _____

49. A _____ is a chord of a circle that contains the center of a circle.

49. _____

50. A _____ is a line that intersects a circle in two points.

50. _____

51. _____ circles are coplanar and have the same center but different radii.

51. _____

52. _____ circles intersect in exactly one point and are coplanar.

52. _____

53. A tangent to a circle is a line that _____ the circle in exactly _____.

53. _____

54. An arc is measured by the measure of its _____ _____.

54. _____

55. A _____ is the set of all points a given distance from a given point.

55. _____

56. The lateral face of a prism is a _____.

56. _____

57. The base of a cone is a _____.

57. _____

58. The base of a pyramid is a _____.

58. _____

59. The bases of prisms are _____.

59. _____

60. The bases of circular cylinders are _____.

60. _____

61. The measure of the diameter of a sphere is _____ its radius.

61. _____

62. A rectangular prism is called a _____.

62. _____

63. A prism having parallelograms as bases is called a _____.

63. _____

64. A polygon with 12 sides is a _____.

64. _____

65. An angle whose vertex is at the center of a circle that is coplanar with the circle is called a _____ _____.

65. _____

Identify each word that is misspelled and write it correctly.

66. Quadralateral

66. _____

67. Isoceles

67. _____

68. Poligonal

68. _____

69. Paralel

69. _____

70. Deameter

70. _____

71. Pyramed

71. _____

72. Enscribed

72. _____

73. Vartex

73. _____

74. Ark

74. _____

75. Suplementery

75. _____

In the four-sided polygon shown, identify each of the following:

76. A pair of adjacent angles

76. _____

77. A pair of adjacent sides

77. _____

78. A diagonal

78. _____

79. A pair of opposite sides

79. _____

80. A pair of opposite angles

80. _____

In the figure shown, line l is parallel to line n, and t is their transversal.
Identify each of the following.

81. A pair of corresponding angles

82. A pair of alternate interior angles

83. A pair of alternate exterior angles

84. A pair of interior angles on the
    same side of the transversal

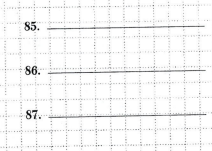

81. _____

82. _____

83. _____

84. _____

Find the indicated sum or difference.

85. $\begin{array}{rrr} 36° & 47' & 15'' \\ +52° & 57' & 59'' \end{array}$

85. _____

86. $\begin{array}{rrr} 46° & 51' & 56'' \\ +21° & 53' & 47'' \end{array}$

86. _____

87. $\begin{array}{rrr} 23° & & \\ -15° & 17' & 42'' \end{array}$

87. _____

88. In the figure shown, $l \parallel n$, $m \parallel p$, and m $\angle 1 = 105°$. Find m $\angle x$.

88. _____

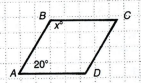

89. In the figure shown, *ABCD* is a parallelogram. Find the measure of
    angle x.

89. _____

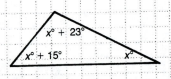

90. In the triangle shown, find the measure of each angle.

90. _____

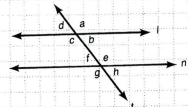

91. If *ABCDEFGHIJ* is a regular decagon, find the measure of angle *HGF*.

91. _____

92. If the diameter of a circle is 15 inches, what is the length of the radius?

92. _____

93. In the figure shown m ∠*AOB* = 124°. Find m ∠*ACB*.

93. _____

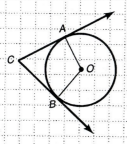

94. In the figure shown, m ∠*AOB* = 110°. Find m ∠*ACB*.

94. _____

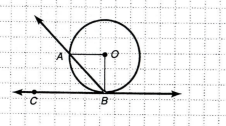

95. In the figure shown, m ∠*AOB* = 90°. Find m ∠*ABC*.

95. _____

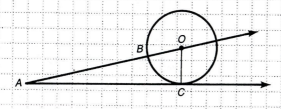

96. In the figure shown, m ∠*AOC* = 70°. Find m ∠*OAC*.

96. _____

Match each of the following to the appropriate drawing.

97. Sphere

98. Hexagonal pyramid

99. Cone

100. Hexagonal prism

101. Triangular pyramid

102. Rectangular parallelepiped

103. Trapezoidal pyramid

104. Cylinder

a.

b.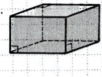

c. 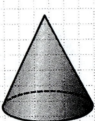 d.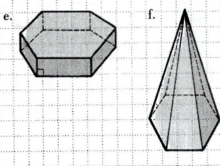

e. 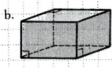 f.

g. h.

97. _____

98. _____

99. _____

100. _____

101. _____

102. _____

103. _____

104. _____

Define each of the following.

ANSWERS

1. Congruent line segments
2. An angle
3. Midpoint
4. A ray
5. Perpendicular lines
6. Acute angle
7. Parallelogram
8. Trapezoid
9. Circle
10. Diameter of a sphere

1. _____
2. _____
3. _____
4. _____
5. _____
6. _____
7. _____
8. _____
9. _____
10. _____

Write each of the following in words.

11. $\angle A$
12. $m\angle A$
13. AB
14. $m\overset{\frown}{PQR}$
15. $\overleftrightarrow{AB} \perp \overleftrightarrow{PQ}$
16. $l \parallel n$
17. $\triangle PQR$
18. $\overrightarrow{RS}$
19. $\overset{\frown}{AB}$
20. $\overleftrightarrow{RT}$

11. _____
12. _____
13. _____
14. _____
15. _____
16. _____
17. _____
18. _____
19. _____
20. _____

Describe one way that each of the following is determined.

21. A line
22. A point
23. A plane
24. Space
25. An angle
26. Explain what is meant by saying that two geometrical figures are congruent.

21. _____
22. _____
23. _____
24. _____
25. _____
26. _____

Find the measure of each unknown in Exercises 27–34.

27.

27. _____

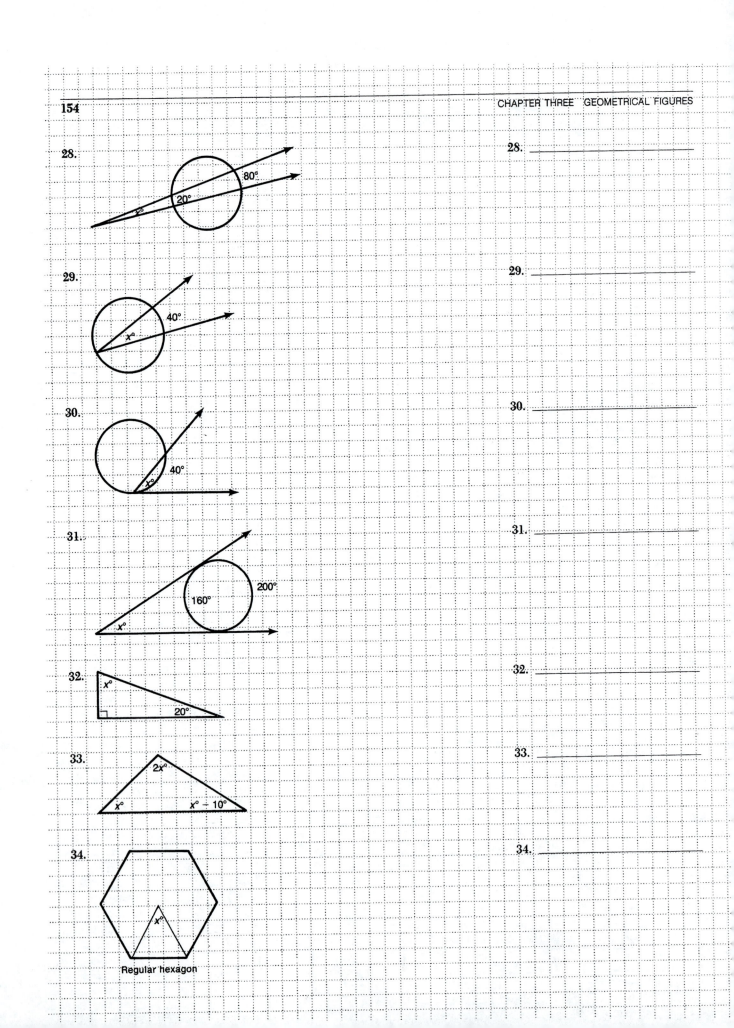

28.

28. _____

29.

29. _____

30.

30. _____

31.

31. _____

32.

32. _____

33.

33. _____

34.

34. _____

Regular hexagon

Match each of the following to the appropriate drawing.

35. Isosceles triangle
36. Regular hexagon
37. Prism
38. Pyramid
39. Cone
40. Plane
41. Half-line
42. Sphere
43. Cylinder
44. Regular triangle
45. Intersecting planes
46. Scalene triangle
47. Equilateral triangle
48. Perpendicular lines
49. Adjacent angles
50. An angle inscribed in a circle.
51. Concentric circles
52. Tangent circles

35. _____
36. _____
37. _____
38. _____
39. _____
40. _____
41. _____
42. _____
43. _____
44. _____
45. _____
46. _____
47. _____
48. _____
49. _____
50. _____
51. _____
52. _____

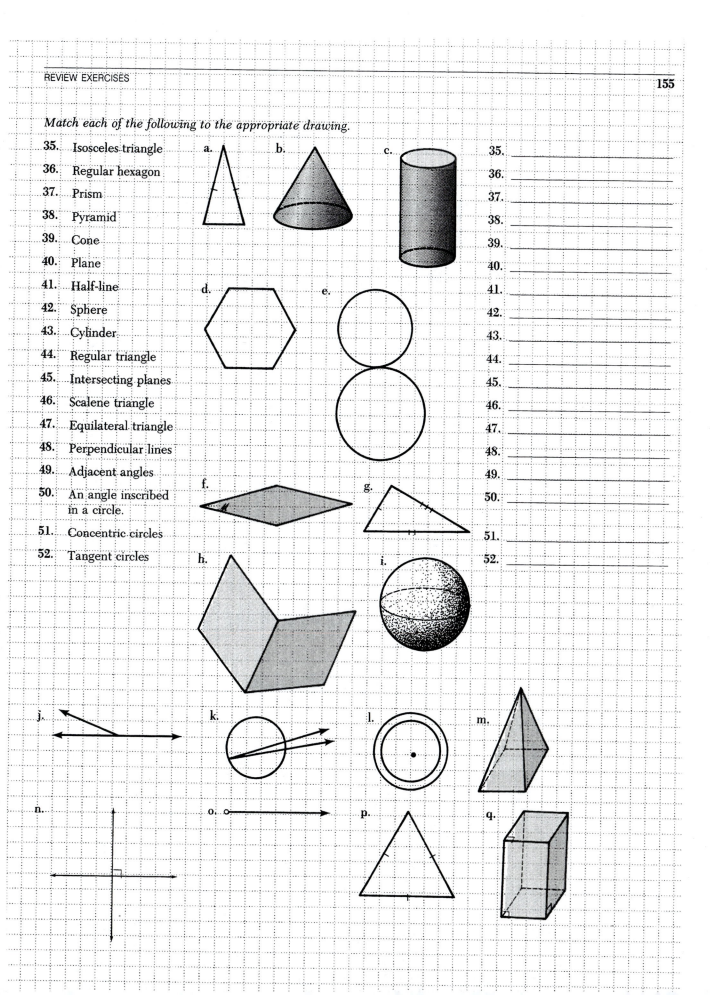

SPECIAL TRIANGLE RELATIONSHIPS; PERIMETER AND CIRCUMFERENCE

4.1 THE PYTHAGOREAN THEOREM

OBJECTIVE

▶ **1** Use the Pythagorean Theorem to solve application problems.

▶ **1** THE PYTHAGOREAN THEOREM

In a right triangle the side opposite the right angle is called the **hypotenuse** and the other two sides are called **legs.**

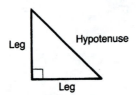

THE PYTHAGOREAN THEOREM

Words: In a right triangle, the sum of the squares of the measures of the legs is equal to the square of the measure of the hypotenuse.

Symbols: In the following triangle, one leg measures *a* units, one leg measures *b* units, and the hypotenuse measures *c* units. The Pythagorean Theorem is written as

$$a^2 + b^2 = c^2$$

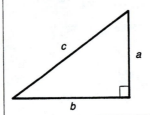

If you are given the length of any two of the three sides in a right triangle, you can find the length of the third side by using the Pythagorean Theorem.

EXAMPLE 1   For each of the following triangles, find the length of the unknown side.

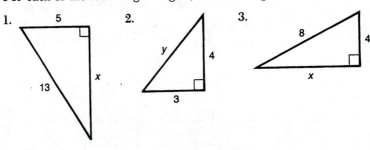

Solution
1. By the Pythagorean Theorem,

$$5^2 + x^2 = 13^2$$
$$25 + x^2 = 169$$
$$x^2 = 169 - 25$$
$$x^2 = 144$$
$$x^2 - 144 = 0$$
$$(x - 12)(x + 12) = 0$$
$$x - 12 = 0 \quad \text{or} \quad x + 12 = 0$$
$$x = 12 \quad \text{or} \quad x = -12$$

Since length cannot be negative, we eliminate the $x = -12$ solution and conclude that the length of the missing side is 12 units.

2. By the Pythagorean Theorem,

$$3^2 + 4^2 = y^2$$
$$9 + 16 = y^2$$
$$25 = y^2$$
$$25 - y^2 = 0$$
$$(5 - y)(5 + y) = 0$$
$$5 + y = 0 \quad \text{or} \quad 5 - y = 0$$
$$y = -5 \quad \text{or} \quad y = 5$$

Since length cannot be negative, $y = 5$ units is the length of the missing side.

3. By the Pythagorean Theorem,

$$x^2 + 4^2 = 8^2$$
$$x^2 + 16 = 64$$
$$x^2 = 48$$
$$x^2 - 48 = 0$$
$$x^2 - (\sqrt{48})^2 = 0$$
$$(x - \sqrt{48})(x + \sqrt{48}) = 0$$
$$x - \sqrt{48} = 0 \quad \text{or} \quad x + \sqrt{48} = 0$$
$$x = \sqrt{48} \quad \text{or} \quad x = -\sqrt{48}$$
$$= 4\sqrt{3} \qquad\qquad = -4\sqrt{3}$$

Since length cannot be negative, $x = 4\sqrt{3}$ units is the length of the missing side. ◀

If we say $\triangle ABC$ has right angle at C, we mean that the vertex of its right angle is at C. It is standard procedure to let a, b, and c denote the lengths of the sides opposite angles A, B, and C, respectively. See Figure 4.1.

FIGURE 4.1

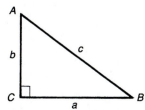

EXAMPLE 2   In $\triangle PQR$ with right angle at P, if $r = 7$ inches and $q = 9$ inches, find p.

Solution   To find the length p, first make a drawing, as shown in the margin.
Then use the Pythagorean Theorem.

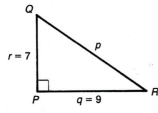

$$p^2 = 7^2 + 9^2$$
$$= 49 + 81$$
$$= 130$$
$$p^2 - (\sqrt{130})^2 = 0$$
$$(p - \sqrt{130})(p + \sqrt{130}) = 0$$
$$p - \sqrt{130} = 0 \qquad \text{or} \qquad p + \sqrt{130} = 0$$
$$p = \sqrt{130} \qquad \text{or} \qquad p = -\sqrt{130}$$

Since length cannot be negative, the length $p = \sqrt{130}$.   ◀

EXAMPLE 3   When a TV is advertised as having a 19-inch screen, it means that the diagonal is 19 inches long. If a 19-inch TV screen has a height of 12 inches, what is the width?

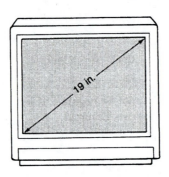

Solution   Let x represent the width of the screen, as shown in the figure.

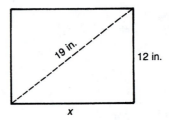

Then by the Pythagorean Theorem,

$$x^2 + 12^2 = 19^2$$
$$x^2 + 144 = 361$$
$$x^2 - 217 = 0$$
$$x^2 - (\sqrt{217})^2 = 0$$
$$(x - \sqrt{217})(x + \sqrt{217}) = 0$$
$$x - \sqrt{217} = 0 \qquad \text{or} \qquad x + \sqrt{217} = 0$$
$$x = \sqrt{217} \qquad \text{or} \qquad x = -\sqrt{217}$$

Since x cannot be negative, the width of the TV is $\sqrt{217}$ inches, or approximately 14.7 inches.   ◀

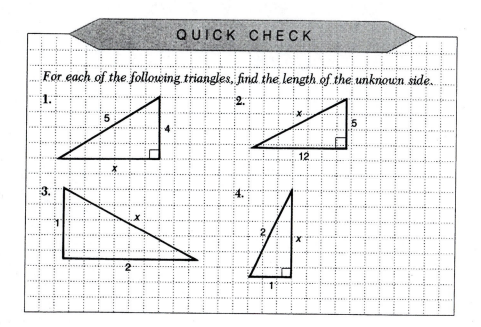

QUICK CHECK

For each of the following triangles, find the length of the unknown side.

1.

2.

3.

4.

A

For each of the following triangles find the length of the unknown side.

ANSWERS

1.

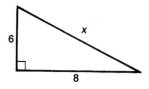

1. _____

2.

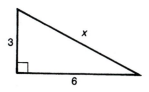

2. _____

3.

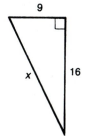

3. _____

4.

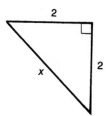

4. _____

5.

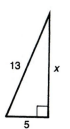

5. _____

6.

6. _____

7.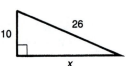

7. _____

8.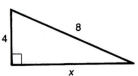

8. _____

9.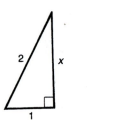

9. _____

10.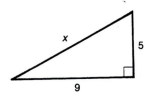

10. _____

Right △ABC has a right angle at C. If the following are the lengths of two sides, find the length of the third side.

11. $a = 6$, $b = 8$

12. $a = 10$, $b = 24$

13. $a = 2$, $c = 4$

14. $a = 3$, $b = 3$

15. $a = 2$, $c = 7$

16. $b = 4$, $c = 9$

17. $a = 2$, $c = 17$

18. $b = 4$, $c = 12$

19. $a = 9$, $b = 20$

20. $a = 4$, $b = 16$

11. _____

12. _____

13. _____

14. _____

15. _____

16. _____

17. _____

18. _____

19. _____

20. _____

21.   In the figure shown, the flag pole is 16 feet tall and perpendicular to the ground. It casts a shadow 6 feet long. How far is it from the top of the pole to the tip of the shadow?

21. _____

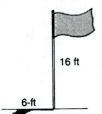

22. The length of the sides of the rectangle shown in the figure are 4 inches and 9 inches. Find the length of its diagonal.

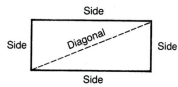

22. _____

23. A boy ties a rope 50 feet long to the top of a tree that is 30 feet tall, pulls the rope tight, and then stakes the end of the rope to the ground. How far is his stake from the bottom of the tree?

23. _____

24. One end of an 8-foot rope is tied to a floating boat at a point on the boat that is at water level. The other end of the rope is tied to the top of a post that is 4 feet above the water. If the rope is taut, how far is the boat from the pier?

24. _____

25. If you walk 3 miles due north and then turn and walk 4 miles due east, how much shorter would your walk have been if you had walked directly from your starting point to your destination?

25. _____

B

For Exercises 26–30, find the unknown length x for each figure.

26.

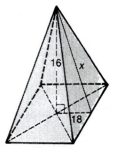

26. _____

27.

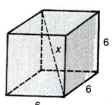

27. _____

28.

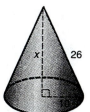

28. _____

29.

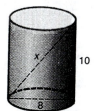

29. _____

30.

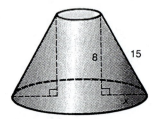

31. Find the slant height of the cone shown in the figure.

31. _____

32. Find the height of the cylinder shown in the figure.

32. _____

33. Find the height of a cone whose radius is 7 inches and whose slant height is 14 inches.

33. _____

34. The area of one face of a triangular prism is 45 square inches. The length of the sides of its right triangular bases is 11 inches. Find the height of the prism.

34. _____

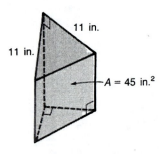

35. Find the height of the pyramid shown in the figure.

35. _____

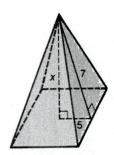

4.2 SPECIAL RIGHT TRIANGLES

OBJECTIVES

1 State the side relationships for a 30°–60°–90° triangle and use these relationships to find side lengths.

2 State the side relationships for a 45°–45°–90° triangle and use these relationships to find side lengths.

3 State the side relationships for 3–4–5 and 5–12–13 right triangles and use these relationships to find side lengths.

1 THE 30°–60°–90° TRIANGLE

An equiangular triangle has three 60° angles. If an altitude is drawn to one side, then the altitude is perpendicular to that side and bisects it. In Figure 4.2 triangle ABC is equiangular and has side length x. $\overline{BM}$ is the altitude to $\overline{AC}$ and bisects $\overline{AC}$, so $AM = (\frac{1}{2})x$. If we let the length of altitude $\overline{BM}$ be y, then we can use the Pythagorean Theorem to find y in terms of the side length x.

$$\left(\frac{1}{2}x\right)^2 + y^2 = x^2$$

$$\frac{1}{4}x^2 + y^2 = x^2$$

$$y^2 = \frac{3}{4}x^2$$

$$y^2 - \frac{3}{4}x^2 = 0$$

$$y^2 - \left(\frac{\sqrt{3}}{2}x\right)^2 = 0$$

$$\left(y - \frac{\sqrt{3}}{2}x\right)\left(y + \frac{\sqrt{3}}{2}x\right) = 0$$

$$y - \frac{\sqrt{3}}{2}x = 0 \qquad \text{or} \quad y + \frac{\sqrt{3}}{2}x = 0$$

$$y = \frac{\sqrt{3}}{2}x \quad \text{or} \qquad\qquad y = -\frac{\sqrt{3}}{2}x$$

FIGURE 4.2

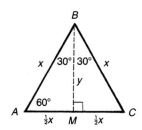

Since length is positive, we can discard the negative answer. Therefore,

$$y = \frac{\sqrt{3}}{2}x$$

In Figure 4.2, triangle *ABM* has angles whose measures are 30°, 60°, and 90°. Notice that y's length is the length of the leg opposite the 30° angle times $\sqrt{3}$. The conclusions we can draw about the side relationships in a **30°–60°–90° triangle** are as follows.

30–60–90 TRIANGLES

1. In a 30–60–90 triangle, the leg opposite the 30° angle is the shortest leg and is one-half the length of the hypotenuse.
2. The length of the leg opposite the 60° angle is the length of the shorter leg times $\sqrt{3}$.
3. The length of the side opposite the 30° angle is the length of the side opposite the 60° angle divided by $\sqrt{3}$.

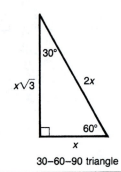

30–60–90 triangle

E X A M P L E 4   For each of the following triangles, find the unknown lengths.

1.   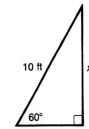   2.   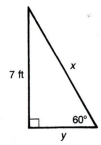   3.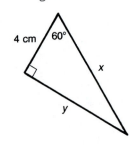

Solution   1. $y = 5$ ft because the side opposite the 30° angle is half the hypotenuse.
$x = 5\sqrt{3}$ ft because the side opposite the 60° angle is the length of the shorter leg times $\sqrt{3}$.

2. $y = \dfrac{7}{\sqrt{3}}$ ft because the side opposite the 30° is the length of the side opposite the 60° angle divided by $\sqrt{3}$.

$x = \dfrac{14}{\sqrt{3}}$ ft because the hypotenuse is twice the length of the side opposite the 30° angle.

3. $x = 8$ cm because the hypotenuse is twice the length of the side opposite the 30° angle.

$y = 4\sqrt{3}$ cm because the side opposite the 60° angle is the length of the shorter leg times $\sqrt{3}$. ◀

QUICK CHECK

In the figure shown triangle ABC has a right angle at C and angle A measures 30°.

1. If $c = 6$ in., find a.

2. If $c = 8$ ft, find b.

3. If $b = 4$ cm, find c.

4. If $b = 3\sqrt{3}$, find a.

5. If $a = 2\sqrt{3}$, find c.

▶2 THE 45°–45°–90° TRIANGLE

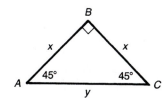

FIGURE 4.3

An isosceles right triangle has two acute 45° angles and is often called a **45–45–90 triangle**. We can use the Pythagorean Theorem to determine the length of the hypotenuse and the side relationships. In Figure 4.3 triangle ABC is an isosceles right triangle with congruent sides of length x and hypotenuse length y. Using the Pythagorean Theorem,

$$x^2 + x^2 = y^2$$
$$2x^2 = y^2$$
$$2x^2 - y^2 = 0$$
$$(\sqrt{2}x)^2 - y^2 = 0$$
$$(\sqrt{2}x - y)(\sqrt{2}x + y) = 0$$
$$\sqrt{2}x - y = 0 \qquad \text{or} \qquad \sqrt{2}x + y = 0$$
$$y = -\sqrt{2}x \qquad \text{or} \qquad y = \sqrt{2}x$$

Since length is positive, we discard the negative answer. Therefore, $y = \sqrt{2}x$.

We can draw the following conclusions about the side relationships in a 45–45–90 triangle.

ANSWERS

1. 3   2. $4\sqrt{3}$ 3. $\dfrac{8}{\sqrt{3}}$ 4. 3   5. $4\sqrt{3}$

45°–45°–90° TRIANGLE

1. In a 45–45–90 triangle, the length of the hypotenuse is the length of a leg times $\sqrt{2}$.
2. The length of a leg is the length of the hypotenuse divided by $\sqrt{2}$.

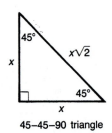

45–45–90 triangle

EXAMPLE 5   For each of the following triangles, find the length of the unknown.

1.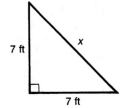

2.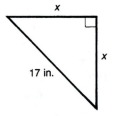

Solution   1. $x = 7\sqrt{2}$ ft because the length of the hypotenuse in a 45–45–90 triangle is the length of a leg times $\sqrt{2}$.

2. $x = \dfrac{17}{\sqrt{2}}$ in. because the length of a leg in a 45–45–90 triangle is the length of the hypotenuse divided by $\sqrt{2}$.   ◀

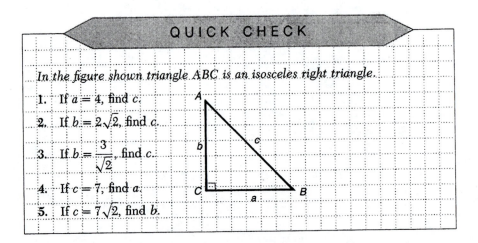

QUICK CHECK

In the figure shown triangle ABC is an isosceles right triangle.

1. If $a = 4$, find c.
2. If $b = 2\sqrt{2}$, find c.
3. If $b = \dfrac{3}{\sqrt{2}}$, find c.
4. If $c = 7$, find a.
5. If $c = 7\sqrt{2}$, find b.

ANSWERS

1. $4\sqrt{2}$ 2. 4 3. 3 4. $\dfrac{7}{\sqrt{2}}$ 5. 7

3-4-5 AND 5-12-13 RIGHT TRIANGLES

Other special right triangles include those whose sides have integer lengths, such as the 3-4-5 right triangle.

The Pythagorean Theorem states how the lengths of the sides of a given right triangle are related. The converse of the Pythagorean Theorem states that if the lengths of the sides of any triangle have a particular relationship, then the triangle must be a right triangle.

CONVERSE OF THE PYTHAGOREAN THEOREM

If a triangle has sides of length a, b, and c and if $a^2 + b^2 = c^2$, then the triangle is a right triangle.

If a, b, and c are integers, then they are called **Pythagorean triples**.

According to the converse of the Pythagorean Theorem, a triangle that has side lengths of 3, 4, and 5 is a right triangle because $3^2 + 4^2 = 5^2$. As a result, the side lengths 3, 4, and 5 are Pythagorean triples. Figure 4.4 shows the 3-4-5 right triangle and two other right triangles whose sides are Pythagorean triples. Note that these other two triangles have side lengths that are multiples of the 3-4-5 triangle side lengths.

FIGURE 4.4
Special right triangles

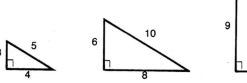
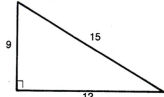

The **5-12-13 right triangle** is another right triangle whose sides are Pythagorean triples because $5^2 + 12^2 = 13^2$. See Figure 4.5.

FIGURE 4.5
The 5-12-13 right triangle

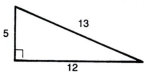

EXAMPLE 6 In the figure shown find the value of the unknown.

Solution The triangle shown is a 5-12-13 right triangle. Thus $x = 12$. ◄

E X A M P L E 7   In the figure shown find the value of the unknown.

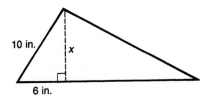

Solution   The triangle is a multiple of the 3–4–5 right triangle. Thus,

$$6 = 2 \cdot 3$$
$$10 = 2 \cdot 5$$
$$x = 2 \cdot 4$$
$$x = 8$$

◀

E X A M P L E 8   In the figure shown find the value of the unknown.

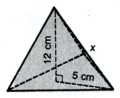

Solution   The triangle is a 5–12–13 right triangle. Thus $x = 13$.

◀

E X A M P L E 9   In the figure shown find the value of the unknown.

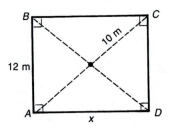

Solution   Since the diagonals of a rectangle bisect each other, $AC = 20$. Then the sides of $\triangle ACD$ are multiples of a 3–4–5 right triangle: $CD = 3 \cdot 4$, $AD = 4 \cdot 4$, and $AC = 5 \cdot 4$. Thus $x = 16$ m.

◀

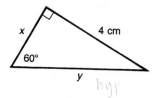

EXERCISE 4.2

A

For Exercises 1–25 find the value of the unknowns.

1.

x   4 cm

60°

y   hyp

2.

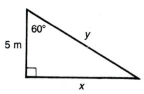

60°

5 m   y

x

3.

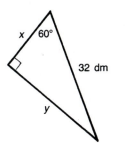

x   60°

32 dm

y

4.

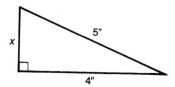

5″

x

4″

5.

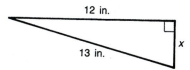

12 in.

x

13 in.

6.

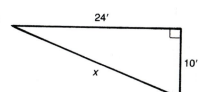

24′

10′

x

7.

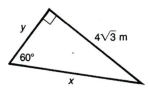

y

$4\sqrt{3}$ m

60°

x

8.

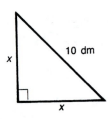

8. _____

9.

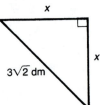

9. _____

10.

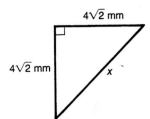

10. _____

11.

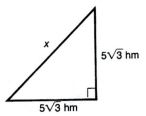

11. _____

12.

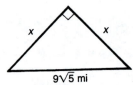

12. _____

13.

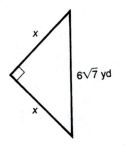

13. _____

14.

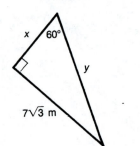

14. _____

15.

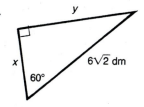

15. _____

16.

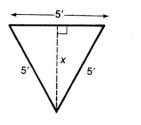

16. _____

17.

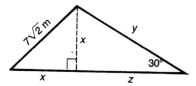

17. _____

18.

18. _____

19.

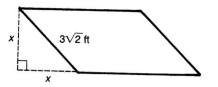

19. _____

20.

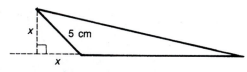

20. _____

21.

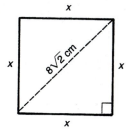

21. _____

22.

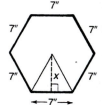

22. _____

23.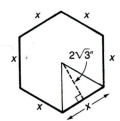

23. _____

24.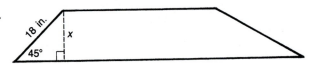

24. _____

25.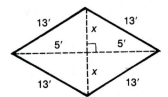

25. _____

B

26. The height of a cone makes an angle of 60° with its slant height. If the radius of the base is 4 inches, find the height and slant height of the cone.

26. _____

27. In the figure shown $\overline{AB}$ and $\overline{CD}$ are diameters and m $\angle ABC = 30°$. Find the radius of the base if $BC = 10$ inches.

27. _____

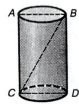

28. The radius of the sphere in the figure is 3000 miles. Find the distance from point A to B.

28. _____

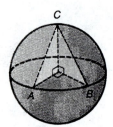

29. The side length of an equilateral rectangular parallelepiped is 5 inches. Find the length of the diagonal of its base.

29. _____

30. The diagonal of a square measures 15 inches. What is its side length?

30. _____

4.3 PERIMETER AND CIRCUMFERENCE

OBJECTIVES

1▶ Define perimeter and find the perimeter of a polygon.

2▶ Define circumference and find the circumference of a circle.

1▶ PERIMETER

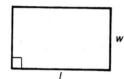

Perimeter is the sum of the lengths of the sides (distance around) of a polygon. For a rectangle whose length is l and whose width is w, the perimeter P is $l + l + w + w$; therefore,

$$P = 2l + 2w$$

The perimeter P of a square whose sides each measure s units is $s + s + s + s$; therefore,

$$P = 4s$$

Thus, to solve perimeter problems, simply determine the variable relationships for the sides or use the given lengths of the sides and then add these lengths to get the perimeter.

EXAMPLE 10   Find the perimeter of the figure shown.

Solution   If we let $x = AB$ and $y = BC$, then $x = 10$ and $y = 5\sqrt{3}$.

$$P = 5 + 10 + 5\sqrt{3}$$
$$= 15 + 5\sqrt{3} \text{ feet}$$

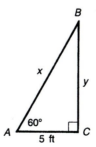

EXAMPLE 11   Find the perimeter of the figure shown.

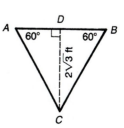

Solution   If we let $x = AD$, then $x = 2$. Thus the sides of the equilateral triangle are each 4 feet.

$$P = 4 + 4 + 4$$
$$= 12 \text{ feet}$$

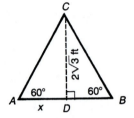

EXAMPLE 12   Find the perimeter.

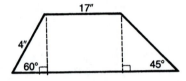

Solution $y = 2$, $z = 2\sqrt{3}$, $w = 2\sqrt{6}$

$$P = 4 + 17 + 2\sqrt{6} + 2\sqrt{3} + 17 + 2$$
$$= 40 + 2\sqrt{6} + 2\sqrt{3} \text{ inches}$$

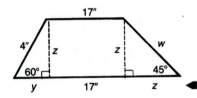

EXAMPLE 13   Find the perimeter of the square shown.

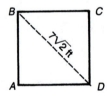

Solution   Since $\triangle ABD$ and $\triangle BCD$ are 45–45–90 triangles, the side lengths are 7 feet; therefore, the perimeter is

$$P = 7 + 7 + 7 + 7$$
$$= 28 \text{ feet}$$

2 ▶ CIRCUMFERENCE

Circumference is the distance around (or length of) a circle. The circumference of a circle with radius r is given by the formula

$$C = 2\pi r$$

In this formula π is the Greek letter pi. Pi is an irrational constant that was first discovered by the Greeks. They noticed that the ratio of the length of a circle to the length of its diameter was always approximately 3.14 units, no matter what size the circle was. See Figure 4.6.

Rounded to the seventh decimal place,

$$\pi = 3.1415927$$

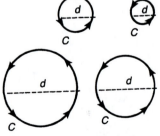

FIGURE 4.6

The ratio of circumference to diameter is always $\pi = C/d$

In this book we use 3.14 as the approximation for π. When a decimal approximation is used for π in a formula, the result is only approximate. For example, suppose the radius of a circle is 2 inches and you want to find its circumference.

$$C = 2\pi r$$
$$= 2(\pi)(2)$$
$$= 4\pi \text{ inches}$$

The exact answer is 4π inches. If π is replaced with its approximate value 3.14, then

$$C \approx 4(3.14) \qquad \text{The symbol } \approx \text{ means "is approximately equal to"}$$
$$\approx 12.56 \text{ inches}$$

Thus, 12.56 inches is the approximate circumference of the circle rounded to two decimal places.

EXAMPLE 14   What is the circumference of a circle whose radius is 18 cm? Use $\pi \approx 3.14$.

Solution
$$C = 2\pi r$$
$$\approx 2(3.14)(18)$$
$$\approx 113.04 \text{ cm}$$
◀

EXAMPLE 15   What is the exact value of the circumference of a circle whose diameter is 15 m?

Solution
$$C = 2\pi r$$
$$= 2(\pi)(7.5)$$
$$= 15\pi \text{ m}$$
◀

EXAMPLE 16   If the circumference of a circle is 6.28 m, find the radius. Use $\pi \approx 3.14$.

Solution
$$C = 2\pi r$$
$$6.28 \approx 2(3.14)r$$
$$\approx 6.28r$$
$$r \approx 1 \text{ m}$$
◀

EXAMPLE 17   If the circumference of a circle is 17π mm, find the diameter.

Solution
$$C = 2\pi r$$
$$17\pi = 2\pi r$$
$$\frac{17\pi}{2\pi} = r$$
$$r = \frac{17}{2}$$

Diameter = 2 · Radius
$$D = 2\left(\frac{17}{2}\right)$$
$$= 17 \text{ mm}$$

Since we never replaced π with a decimal approximation, this value for the diameter is exact.
◀

QUICK CHECK

1. What is the perimeter of a regular hexagon whose sides measure 3.2 inches?

2. What is the perimeter of an isosceles triangle whose congruent sides measure 4 inches each and whose third side measures 3 inches?

3. In a 30–60–90 triangle, the hypotenuse measures 8 inches. Find the perimeter.

4. In a 45–45–90 triangle, the hypotenuse measures 10 cm. Find the perimeter.

5. The diagonals of a rhombus measure 24 in. and 32 in. Find the perimeter.

ANSWERS

1. 19.2 in. **2.** 11 in. **3.** $12 + 4\sqrt{3}$ in. **4.** $10 + 10\sqrt{2}$ cm **5.** 80 in.

EXERCISE 4.3

A

Find the perimeter or circumference for each of the following geometrical figures.

ANSWERS

1.

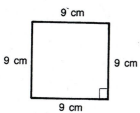

1. _____

2.

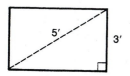

2. _____

3.

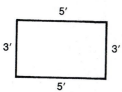

3. _____

4.

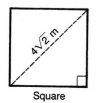

Square

4. _____

5.

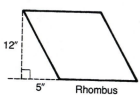

Rhombus

5. _____

6.

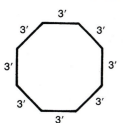

6. _____

7.

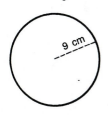

7. _____

8.

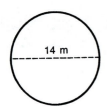

14 m

8. _____

9.

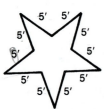

5′ 5′
5′   5′
5′   5′
5′   5′
5′ 5′

9. _____

10.

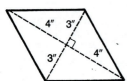

4″ 3″
3″   4″

10. _____

11.

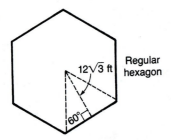

12√3 ft   Regular
hexagon

60°

11. _____

12.

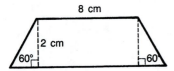

8 cm

2 cm

60°   60

12. _____

13.

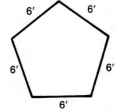

6′   6′

6′   6′

6′

13. _____

14.

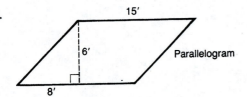

15′

6′

Parallelogram

8′

14. _____

15.

1 ft 3 in.

15. _____

16.

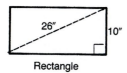

26″   10″

Rectangle

16. _____

17.

17 ft

45°   45°

17. _____

18.

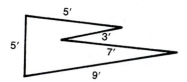

5′

5′   3′

7′

9′

18. _____

19.

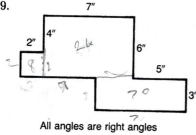

7″

4″

2″

6″

5″

3″

All angles are right angles

19. _____

20.

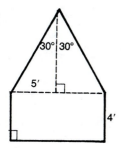

30° 30°

5′

4′

20. _____

21.

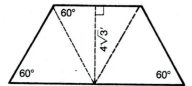

60°

$4\sqrt{3}'$

60°   60°

21. _____

22.

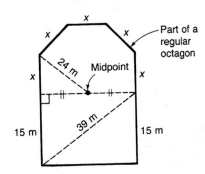

22. _____

23. The circumference of a circle is 16 inches. Find the radius.

24. The circumference of a circle is 6π. Find the diameter.

25. The circumference of a tree trunk is 39 inches. Find the radius.

26. The base of a cylinder has a circumference of 3π feet. Find the diameter.

27. The base of a cylinder has a circumference of 7π feet. Find the diameter.

28. The base of a cone has a circumference of 7π inches and the height of the cone is 12 inches. Find the slant height.

29. The base of a cone has a circumference of 9π inches and the height of the cone is 25 inches. Find the slant height.

30. What is the circumference of a 9-inch pie pan?

23. _____

24. _____

25. _____

26. _____

27. _____

28. _____

29. _____

30. _____

B

31. Find the perimeter of a square that has a side of 6 inches.

32. Find the perimeter of a square whose diagonal measures $32\sqrt{2}$ centimeters.

33. A farmer has 100 feet of fencing material and plans to build two adjacent rectangular pens. If the width of the pens is to be 12 feet, what must be the total length of the two pens?

34. A running track is shaped as shown in the figure. The ends are semicircles. What is the total length of the track?

31. _____

32. _____

33. _____

34. _____

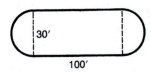

ANSWERS

1.   In the triangle shown, $\overline{BD}$ is the altitude to $\overline{AC}$. Find BC.

1. _____

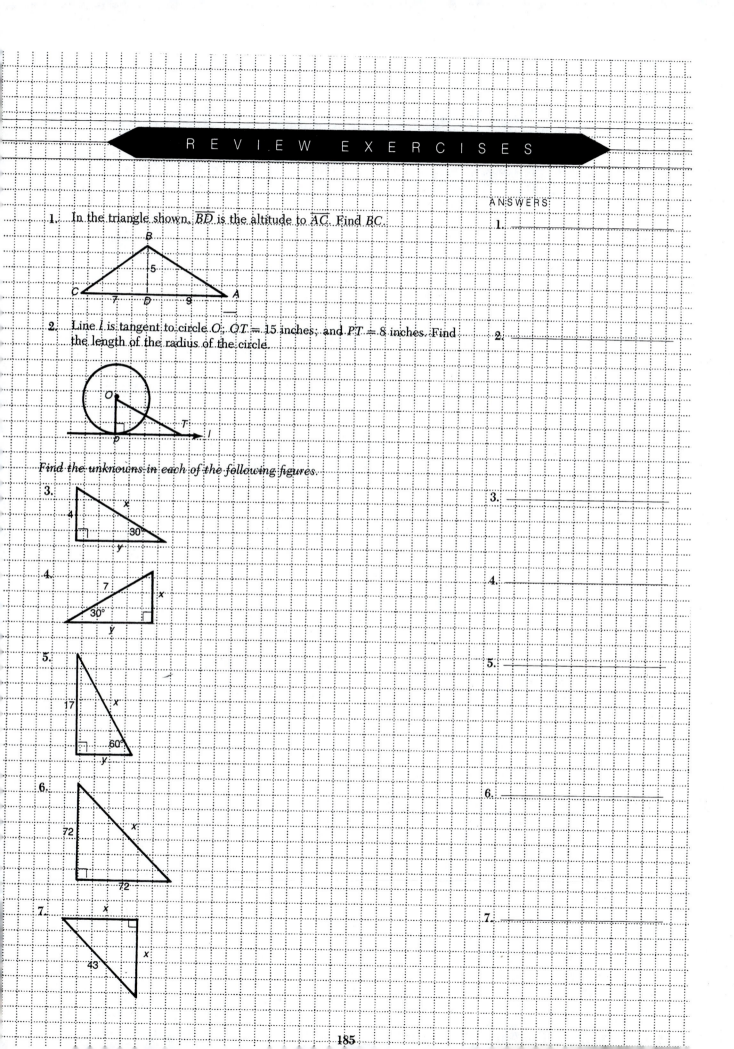

2.   Line l is tangent to circle O; $OT = 15$ inches; and $PT = 8$ inches. Find the length of the radius of the circle.

2. _____

Find the unknowns in each of the following figures.

3.

3. _____

4.

4. _____

5.

5. _____

6.

6. _____

7.

7. _____

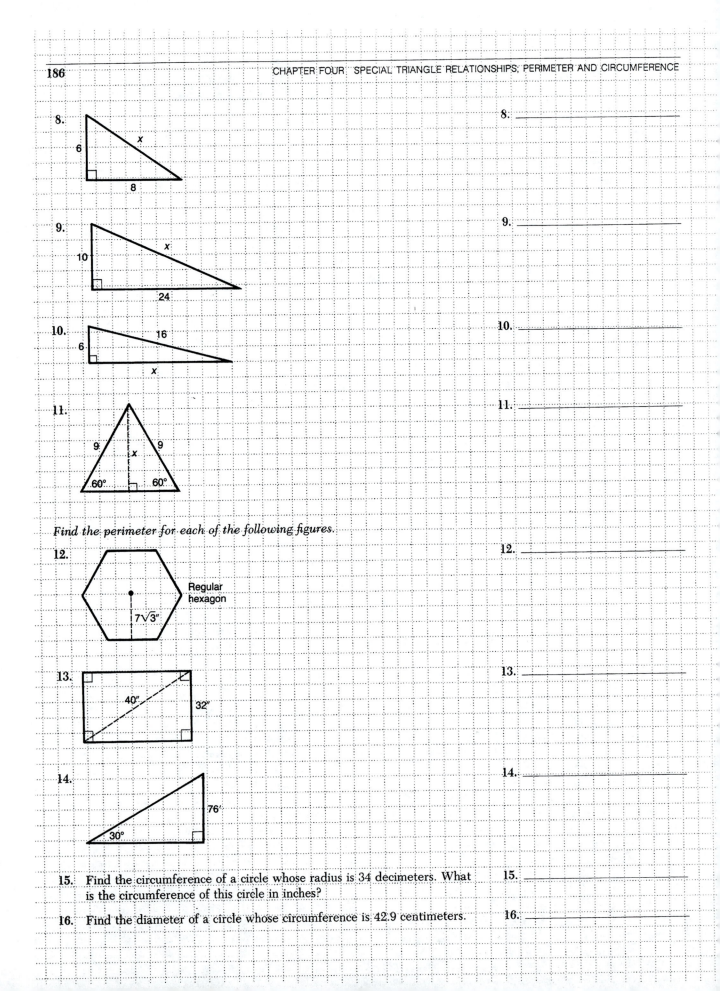

8. _____

9. _____

10. _____

11. _____

Find the perimeter for each of the following figures.

12. _____

13. _____

14. _____

15. Find the circumference of a circle whose radius is 34 decimeters. What   15. _____
    is the circumference of this circle in inches?

16. Find the diameter of a circle whose circumference is 42.9 centimeters.   16. _____

ANSWERS

Find the unknowns in each of the following triangles.

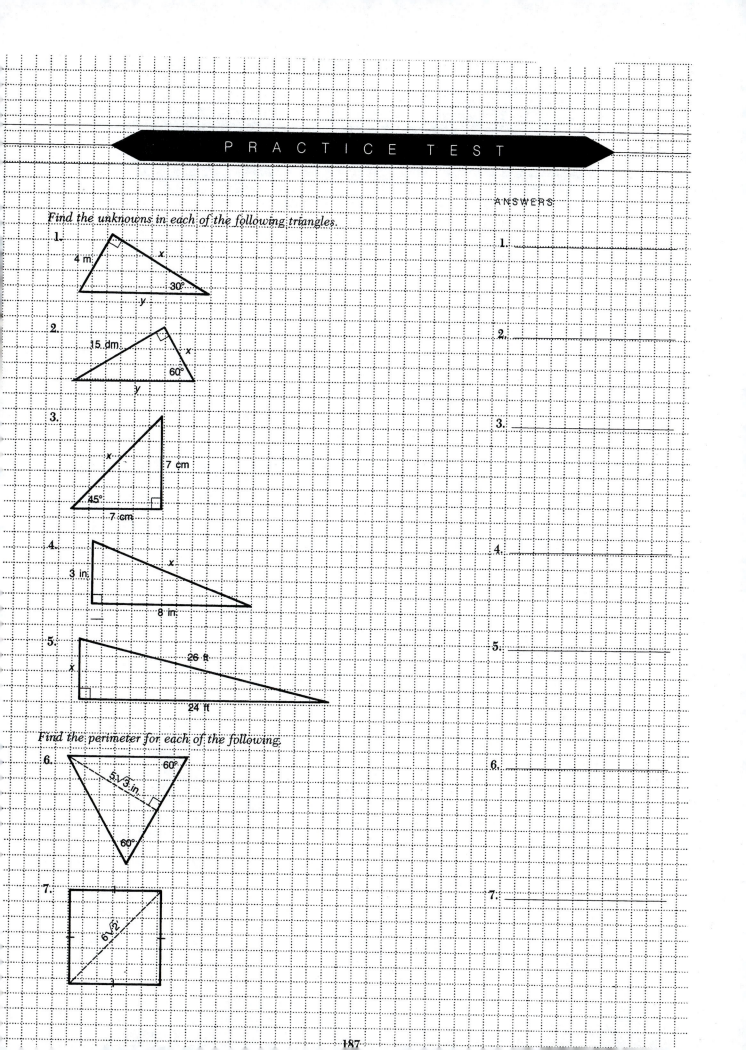

1.

4 m   x   30°   y

2.

15 dm   x   60°   y

3.

x   7 cm   45°   7 cm

4.

3 in.   x   8 in.

5.

26 ft   x   24 ft

Find the perimeter for each of the following.

6.

60°   5√3 in.   60°

7.

6√2

1. _____

2. _____

3. _____

4. _____

5. _____

6. _____

7. _____

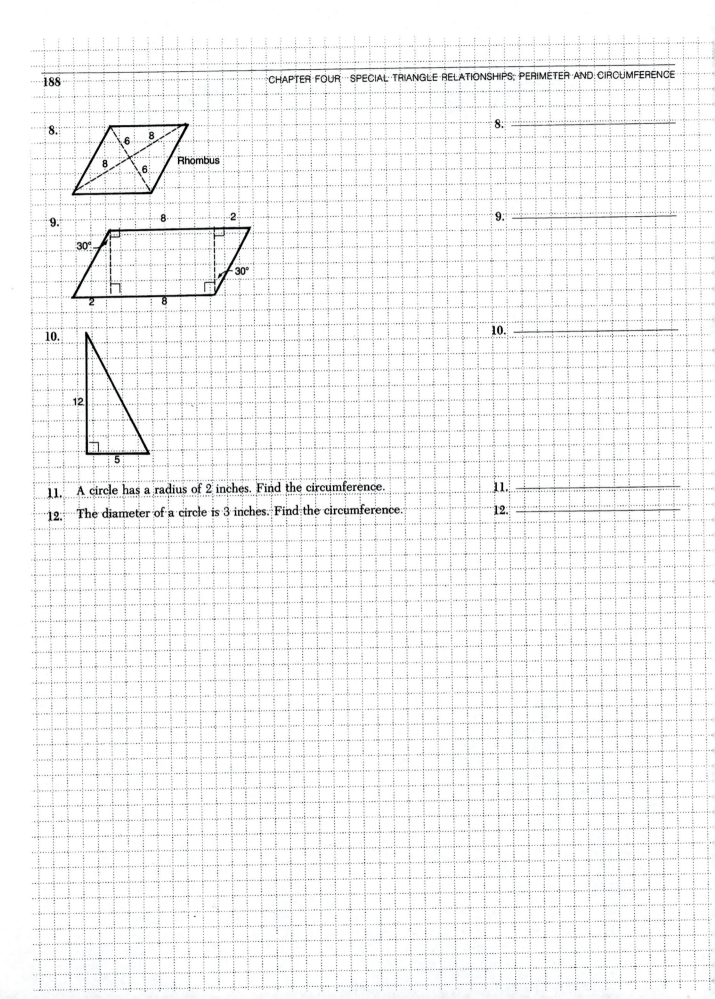

8.   8. _____

9.   9. _____

10.   10. _____

11. A circle has a radius of 2 inches. Find the circumference.   11. _____

12. The diameter of a circle is 3 inches. Find the circumference.   12. _____

CHAPTER FIVE

AREA
AND
VOLUME

5.1 AREA

OBJECTIVES

1. ▶ Find the area of a rectangle and a square; convert square units.
2. ▶ Find the area of a triangle.
3. ▶ Find the area of a parallelogram.
4. ▶ Find the area of a trapezoid.
5. ▶ Find the area of a hexagon.
6. ▶ Find the area of a circle.

1 ▶ AREA OF A RECTANGLE AND A SQUARE

Area is a measure of surface. If you want to paint a room, you need to determine how much wall surface there is to know how much paint to buy. Similarly, when carpeting a room, you need to know the area of the floor to determine the amount of carpet to buy.

Area is measured in **square units**. Figure 5.1 shows the relative size of a square inch, a square centimeter, and a square decimeter.

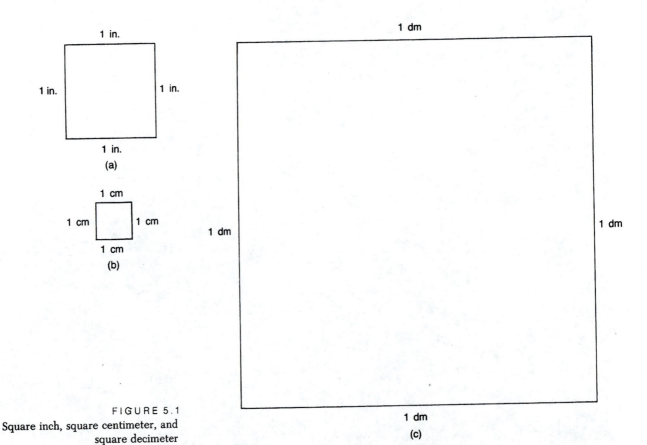

FIGURE 5.1
Square inch, square centimeter, and square decimeter

To find the area of a plane geometrical figure means to find how many square units it takes to cover the surface enclosed by the figure. Figure 5.2 shows rectangle *ABCD*. Notice that it takes 22 square centimeters to cover the surface enclosed by the rectangle.

FIGURE 5.2

A rectangle whose area is 22 square centimeters

B | | | | | | | | | | | *C*

1	2	3	4	5	6	7	8	9	10	11
12	13	14	15	16	17	18	19	20	21	22

A | | | | | | | | | | | *D*

Square units are abbreviated by preceding the unit name with "sq" or by placing the exponent 2 on the unit abbreviation. For example,

Square inch: sq in. or in.^{2}

Square feet: sq ft or ft^2

Square centimeter: sq cm or cm^2

Square meter: sq m or m^2

To find the area of the rectangle in Figure 5.2, multiply its length, 11 centimeters, by its width, 2 centimeters. The resulting area is 22 sq cm or 22 cm^2. Multiplying length times width simply counts the number of square units the rectangle contains.

AREA OF A RECTANGLE

$$A = l \cdot w$$

Since a square is a rectangle, its area can be obtained by multiplying its length by its width. However, because its length and width are the same, this is equivalent to squaring the side length.

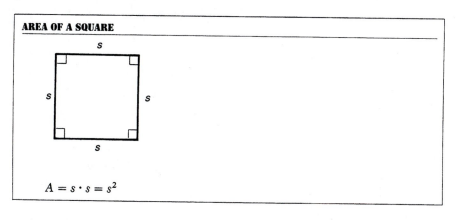

AREA OF A SQUARE

$$A = s \cdot s = s^2$$

EXAMPLE 1   Find the area of the rectangle shown.

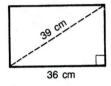

Solution   To determine the area of this rectangle, first we need to find the width. Let x be the width and use the Pythagorean Theorem.

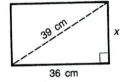

$$x^2 + 36^2 = 39^2$$
$$x^2 + 1296 = 1521$$
$$x^2 = 1521 - 1296$$
$$= 225$$
$$x = 15 \text{ cm}$$

Now that we know both the length and width of the rectangle, we can calculate the area.

$$A = l \cdot w$$
$$= 15 \cdot 36$$
$$= 540 \text{ cm}^2$$ ◀

EXAMPLE 2   Find the area of the square shown.

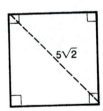

Solution   Since the figure is a square, all of the sides have the same length. Let s be the length of a side and use the Pythagorean Theorem.

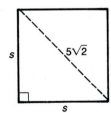

$$s^2 + s^2 = (5\sqrt{2})^2$$
$$2s^2 = 50$$
$$s^2 = 25$$
$$s = 5$$

Now that we know the length of one side of the square, we can find the area.

$$\text{Area} = s^2$$
$$= 5^2$$
$$= 25 \text{ in.}^2$$ ◀

When dealing with problems involving area, we often need to convert from one unit to another. To do this we use the square of a unit conversion factor. For example,

$$\left(\frac{1 \text{ yd}}{3 \text{ ft}}\right)^2 = \frac{1 \text{ yd}}{3 \text{ ft}} \cdot \frac{1 \text{ yd}}{3 \text{ ft}}$$
$$= \frac{1 \text{ yd}^2}{9 \text{ ft}^2}$$

Suppose you wanted to carpet a room that was 16 feet wide and 20 feet long. Since carpet is sold by the square yard, you need to know how many square yards it would take to cover your floor. A square yard, abbreviated sq yd or yd², is the measure of the surface of a square 1 yard by 1 yard. Figure 5.3 shows a room 16 feet by 20 feet that has been divided into square yards. Notice that each square yard contains 9 square feet. A square foot, abbreviated sq ft or ft², is the measure of a square 1 foot by 1 foot.

FIGURE 5.3

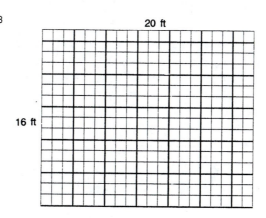

The room pictured in Figure 5.3 contains 30 square yards plus 50 square feet. Since 1 sq yd = 9 sq ft, the 50 square feet can be divided into square yards by dividing 50 by 9.

$$50 \text{ sq ft} = \frac{50}{9} \text{ sq yd}$$

$$= 5\frac{5}{9} \text{ sq yd}$$

Thus the total amount of carpet needed is

$$30 \text{ sq yd} + 5\frac{5}{9} \text{ sq yd} = 35\frac{5}{9} \text{ sq yd}$$

As shown in Example 3, another way to approach this problem is to use the formula for finding the area of a rectangle ($A = l \cdot w$) and then convert the resulting units to square yards.

EXAMPLE 3 Find the area of a room 16 feet by 20 feet in square yards.

Solution First use the formula for finding the area of a rectangle.

$$A = l \cdot w$$
$$= 20 \text{ ft} \cdot 16 \text{ ft}$$
$$= 320 \text{ ft}^2$$

Now convert 320 square feet to square yards.

$$320 \text{ ft}^2 = 320 \text{ ft}^2 \left(\frac{1 \text{ yd}}{3 \text{ ft}}\right)^2 \qquad \text{Using the same unit relationships but squaring them}$$

$$= 320 \text{ ft}^2 \cdot \frac{1 \text{ yd}^2}{9 \text{ ft}^2} \qquad \text{Squaring the numerator and denominator}$$

$$= 35\frac{5}{9} \text{ yd}^2 \qquad \blacktriangleleft$$

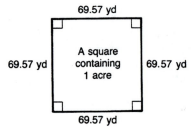

FIGURE 5.4
One acre

Area units can be the square of any American or metric unit of length. In the American system the area unit for land measure is the **acre** (abbreviated ac)

$$1 \text{ acre} = 4840 \text{ yd}^2$$
$$1 \text{ acre} = 43{,}560 \text{ ft}^2$$

A square whose measure is 1 acre is approximately 70 yards by 70 yards. See Figure 5.4.

A square mile contains 640 acres and is called a **section** of land.

$$1 \text{ mi}^2 = 640 \text{ acres} = 1 \text{ section}$$

In the metric system the fundamental unit for measuring land area is the **are** (pronounced "air" and abbreviated a). Although the metric system has seven units for land measure, the **hectare** (abbreviated ha) is used most often. A square hectare is 100 meters on a side; that is, 1 hectare = 10,000 square meters. See Figure 5.5.

FIGURE 5.5

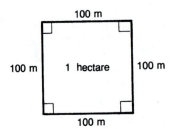

The link between the American and metric systems for area is the relationship between the hectare and the acre. See Figure 5.6.

FIGURE 5.6
1 hectare = 2.47 acres

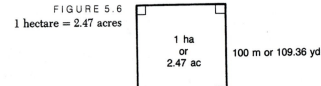

EXAMPLE 4   Perform the following unit conversions.

1. $236 \text{ in.}^2 = \underline{\hspace{1cm}} \text{ ft}^2$ 2. $2354 \text{ mm}^2 = \underline{\hspace{1cm}} \text{ cm}^2$
3. $54 \text{ ft}^2 = \underline{\hspace{1cm}} \text{ dm}^2$ 4. $3 \text{ ac} = \underline{\hspace{1cm}} \text{ ha}$

Solution   1. $236 \text{ in.}^2 = 236 \text{ in.}^2 \left(\dfrac{1 \text{ ft}}{12 \text{ in.}} \right)^2$

$$= 236 \text{ in.}^2 \cdot \frac{1 \text{ ft}^2}{144 \text{ in.}^2}$$

$$= 1.64 \text{ ft}^2$$

2. $2354 \text{ mm}^2 = 2354 \text{ mm}^2 \left(\dfrac{1 \text{ cm}}{10 \text{ mm}} \right)^2$

$$= 2354 \text{ mm}^2 \cdot \frac{1 \text{ cm}^2}{100 \text{ mm}^2}$$

$$= 23.54 \text{ cm}^2$$

3. $54 \text{ ft}^2 = 54 \text{ ft}^2 \left(\dfrac{12 \text{ in.}}{1 \text{ ft}}\right)^2 \left(\dfrac{2.54 \text{ cm}}{1 \text{ in.}}\right)^2 \left(\dfrac{1 \text{ dm}}{10 \text{ cm}}\right)^2$

$\qquad = 54 \text{ ft}^2 \cdot \dfrac{144 \text{ in.}^2}{1 \text{ ft}^2} \cdot \dfrac{6.45 \text{ cm}^2}{1 \text{ in.}^2} \cdot \dfrac{1 \text{ dm}^2}{100 \text{ cm}^2}$

$\qquad = 501.55 \text{ dm}^2$

4. $3 \text{ ac} = 3 \text{ ac} \cdot \dfrac{1 \text{ ha}}{2.47 \text{ ac}}$

$\qquad = 1.21 \text{ ha}$ ◀

QUICK CHECK

1. Find the area of a rectangle whose sides measure 5 inches and 17.3 inches.

2. Find the area of a rectangle whose sides measure 8 feet and 19 feet.

3. A rectangle has an area of 70 square inches. If the length is 14.8 inches, find the width.

4. Find the area of a square whose sides measure 1.6 inches.

5. Find the area of a square whose sides measure 3.06 decimeters.

6. The diagonal of a square measures 16 inches. Find the area.

Perform the following unit conversions.

7. $15 \text{ ft}^2 =$ _____ in.^2 8. $25 \text{ m}^2 =$ _____ dm^2

9. $12 \text{ in.}^2 =$ _____ cm^2 10. $6 \text{ ha} =$ _____ ac

11. A rectangle is 7 feet by 6 feet. Find its area in square inches.

▶ **2**   AREA OF A TRIANGLE

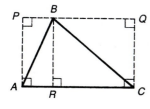

FIGURE 5.7

To find the area of triangle ABC in Figure 5.7, we first draw a rectangle $APQC$ that has length equal to one side of the triangle (in our case AC) and that has width equal to the altitude from the opposite angle (in our case $\angle ABC$). Now let's make some observations about the area of the resulting figure.

1. The area of $\triangle ABC$ = Area of $\triangle ABR$ + Area of $\triangle BRC$.
2. The area of $\triangle ABR$ is $\frac{1}{2}$ the area of rectangle $APBR$.
3. The area of $\triangle BRC$ is $\frac{1}{2}$ the area of rectangle $BQCR$.

ANSWERS

1. 86.5 in.^{2} 2. 152 ft^2 3. 4.73 in.   4. 2.56 in.^{2} 5. 9.36 in.^{2} 6. 128 in.^{2} 7. 2160 in.^{2}
8. 2500 dm^2 9. 77.42 cm^2 10. 14.82 ac   11. 6048 in.^{2}

$$\text{Area } \triangle ABC = \left(\frac{1}{2}\right)(AR \cdot AP) + \left(\frac{1}{2}\right)(RC \cdot QC)$$

$$= \left(\frac{1}{2}\right)(AR)(AP) + \left(\frac{1}{2}\right)(RC)(AP) \quad \text{Since } AP = QC$$

$$= \left(\frac{1}{2}\right)(AP)(AR + RC)$$

$$= \left(\frac{1}{2}\right)(AP)(AC) \qquad \text{By the Definition of}$$
$$\text{Between, } AR + RC = AC$$

In Figure 5.7, AP is the length of the altitude to $\overline{AC}$. If we designate the length of an altitude as the height of a triangle and the side to which the altitude is drawn as its base, then the area of the triangle is one-half its base times its height.

AREA OF A TRIANGLE

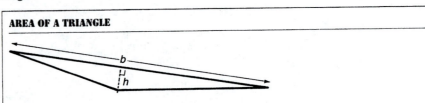

Words: The area of a triangle is one-half the length of a side multiplied by the length of the altitude drawn to the side.

Symbols: $A = \dfrac{1}{2}(b \cdot h)$

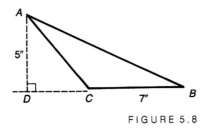

FIGURE 5.8

When the altitude falls outside the triangle, you still multiply the length of the altitude by the length of the side to which it was drawn. In Figure 5.8, $\overline{AD}$ is the altitude to $\overline{BC}$ and is considered the height of the triangle when $\overline{BC}$ is the base, so the area is calculated as

$$A = \frac{1}{2}(5)(7)$$

$$= 17.5 \text{ sq in.}$$

When determining the area of a right triangle, if either leg is chosen as the base, then the other leg is the altitude to that base or the height of the triangle.

AREA OF A RIGHT TRIANGLE

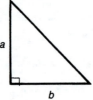

Words: The area of a right triangle is equal to one-half the product of the length of its legs.

Symbols: $A = \dfrac{1}{2}(a \cdot b)$

To find the area of an equilateral triangle, we use the side and altitude relationships we developed in Section 4.2. The formula for finding the area of an equilateral triangle is shown below. If you choose not to learn this formula, you can find the area of an equilateral triangle by using the 30–60–90 side relationships to find its altitude or side length and then use the formula $A = \frac{1}{2}(b \cdot h)$.

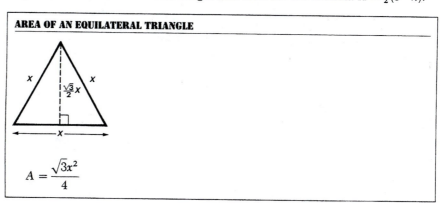

AREA OF AN EQUILATERAL TRIANGLE

$$A = \frac{\sqrt{3}x^2}{4}$$

EXAMPLE 5   Find the area of the triangle shown.

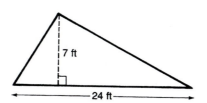

Solution
$$A = \frac{1}{2}(b \cdot h)$$
$$= \left(\frac{1}{2}\right)(24)(7)$$
$$= 12 \cdot 7$$
$$= 84 \text{ ft}^2$$

EXAMPLE 6   Find the area of the triangle shown.

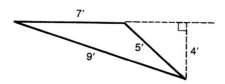

Solution
$$A = \frac{1}{2}(b \cdot h)$$
$$= \left(\frac{1}{2}\right)(7)(4)$$
$$= 14 \text{ ft}^2$$

EXAMPLE 7   Find the area of the triangle shown.

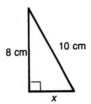

Solution   Use the Pythagorean Theorem to find the length of the leg.

$$x^2 + 8^2 = 10^2$$
$$x^2 + 64 = 100$$
$$x^2 = 100 - 64$$
$$= 36$$
$$x = 6$$

Therefore,

$$A = \left(\frac{1}{2}\right)(6)(8)$$
$$= 24 \text{ cm}^2$$

EXAMPLE 8   Find the area of the triangle shown.

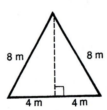

Solution   Let h be the height shown. By the Pythagorean Theorem,

$$h^2 + 4^2 = 8^2$$
$$h^2 + 16 = 64$$
$$h^2 = 48$$
$$h = \sqrt{48}$$
$$= 4\sqrt{3} \text{ m}$$

Then since the side to which the height is drawn measures $4 + 4 = 8$ m, it follows that

$$A = \left(\frac{1}{2}\right)(8)(4\sqrt{3})$$
$$= 16\sqrt{3} \text{ m}^2$$

EXAMPLE 9   Find the area of the triangle shown.

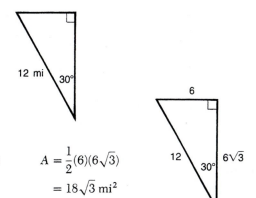

Solution

$$A = \frac{1}{2}(6)(6\sqrt{3})$$
$$= 18\sqrt{3} \text{ mi}^2$$

QUICK CHECK

1. Find the area of a triangle that has a base measuring 7 inches and height measuring 5 inches.

2. Find the area of a 30–60–90 right triangle whose hypotenuse is 26 centimeters.

3. If the hypotenuse of a 45–45–90 right triangle is 16 inches, find the area.

4. What is the area of a 3–4–5 right triangle?

5. What is the area of a 5–12–13 right triangle?

6. An equilateral triangle has an altitude of $4\sqrt{3}$ inches. What is the area?

❸ ▶ AREA OF A PARALLELOGRAM

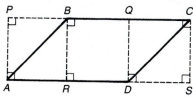

FIGURE 5.9

The formula for the area of a parallelogram is obtained in much the same way as the triangle formula. Consider the parallelogram $ABCD$ in Figure 5.9. Rectangle $APCS$ is drawn having length AS and width equal to the height of the parallelogram. (The height of a parallelogram is the distance between a pair of parallel sides.)

The area of parallelogram $ABCD$ is equal to

area of rectangle $BQDR$ + area of $\triangle ABR$ + area of $\triangle QCD$

Since triangles ABR and QCD are the same size, the area of parallelogram $ABCD$ is equal to

area of rectangle $BQDR$ + 2 · (area of $\triangle ABR$)

ANSWERS

1. 17.5 sq in.   2. (84.5)$\sqrt{3}$ sq cm   3. 64 sq in.   4. 6   5. 30   6. 16$\sqrt{3}$ sq in.

$$\text{Area of parallelogram } ABCD = BQ \cdot BR + 2\left[\frac{1}{2}(AR \cdot BR)\right]$$

$$= BQ \cdot BR + AR \cdot BR$$

$$= BR(BQ + AR)$$

$$= BR(RD + AR) \qquad \text{Since } BQ = RD$$

$$= BR(AD)$$

Notice that BR (the height) is the distance between the parallel sides $\overline{AD}$ and $\overline{BC}$ and that AD is the length of one of these parallel sides. Thus, the area of a parallelogram is equal to the product of the length of a side and the height drawn to that side.

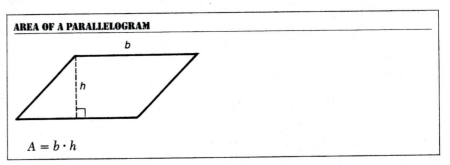

AREA OF A PARALLELOGRAM

$$A = b \cdot h$$

The area of a rhombus can be determined based on the fact that its diagonals are perpendicular and bisect each other.

In Figure 5.10, the area of rhombus $ABCD$ is equal to the area of $\triangle AOD$ + area of $\triangle AOB$ + area of $\triangle BOC$ + area of $\triangle DOC$. The area of each of these right triangles is $\frac{1}{2}(ab)$; therefore, the area of the rhombus is

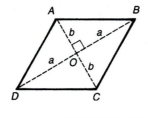

FIGURE 5.10

$$\text{Area of rhombus } ABCD = 4\left[\frac{1}{2}(ab)\right]$$

$$= 2ab$$

A more convenient way to arrange this area formula is

$$\text{Area of rhombus } ABCD = \frac{1}{2}(2a \cdot 2b)$$

In this formula $2a$ is the length of one diagonal and $2b$ is the length of the other diagonal. If the diagonal lengths are designated d_1 and d_2, then the area of the rhombus is one-half the product of the length of its diagonals.

AREA OF A RHOMBUS

If $AC = d_1$ and $BD = d_2$, then

$$A = \frac{1}{2}(d_1 \cdot d_2)$$

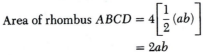

EXAMPLE 10   Find the area of the parallelogram shown.

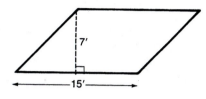

Solution $A = b \cdot h$

$= (15)(7)$

$= 105 \text{ ft}^2$

EXAMPLE 11   Find the area of the parallelogram shown.

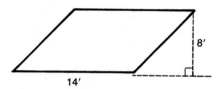

Solution $A = b \cdot h$

$= (14)(8)$

$= 112 \text{ ft}^2$

EXAMPLE 12   Find the area of the parallelogram shown.

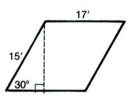

Solution $A = b \cdot h$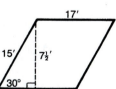

$= (17)(7.5)$

$= 127.5 \text{ ft}^2$

EXAMPLE 13   Find the area of the rhombus shown.

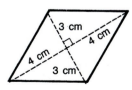

Solution $A = \frac{1}{2}(d_1 \cdot d_2)$

$= \frac{1}{2}(6 \cdot 8)$

$= 24 \text{ cm}^2$

QUICK CHECK

1. A parallelogram has a base of 3 inches and a height of 5 inches. What is the area?

2. One angle of a parallelogram measures 60° and the sides measure 8 inches and 10 inches. Find the area.

3. If the diagonals of a rhombus measure 5 inches and 8 inches, find the area.

4. The sides of a rhombus measure 10 meters and one diagonal measures 16 meters. Find the area of the rhombus.

 AREA OF A TRAPEZOID

To find the area of a trapezoid consider trapezoid $ABCD$ in Figure 5.11. Two heights are drawn to divide the trapezoid's area into three parts. The area of trapezoid $ABCD$ is equal to the area of $\triangle ABP$ + area of rectangle $BCQP$ + area of $\triangle CQD$.

FIGURE 5.11

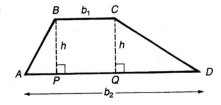

$$\text{Area of } \triangle APB = \frac{1}{2}(AP)(h)$$

$$\text{Area of rectangle } BCQP = b_1 h$$

$$\text{Area of } \triangle CQD = \frac{1}{2}(QD)(h)$$

$$\text{Area of trapezoid } ABCD = \frac{1}{2}(AP)(h) + b_1 h + \frac{1}{2}(QD)(h)$$

$$= \frac{1}{2}h(AP + QD + 2b_1)$$

$$= \frac{1}{2}h(AP + QD + b_1 + b_1)$$

$$= \frac{1}{2}h(AP + QD + PQ + b_1) \qquad \text{Since } b_1 = PQ$$

$$= \frac{1}{2}h(b_1 + b_2) \qquad \text{Since } AP + PQ + QD = b_2$$

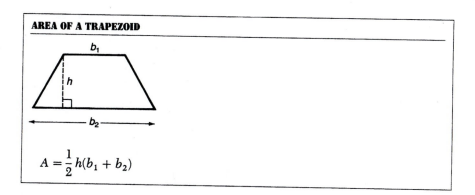

AREA OF A TRAPEZOID

$$A = \frac{1}{2} h(b_1 + b_2)$$

E X A M P L E 1 4   Find the area of the trapezoid shown.

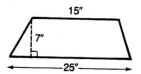

Solution $A = \frac{1}{2} h(b_1 + b_2)$

$\quad\quad = \frac{1}{2} (7)(15 + 25)$

$\quad\quad = \frac{1}{2} (7)(40)$

$\quad\quad = 140 \text{ in.}^2$

◄

E X A M P L E 1 5   Find the area of the trapezoid shown.

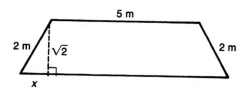

Solution   We know that the height h of the trapezoid is $\sqrt{2}$ m and that its top base, b_1, is 5 m. To calculate the area of the trapezoid, we must also know the bottom base, b_2. First, we find the length x by using the Pythagorean Theorem.

$$x^2 + (\sqrt{2})^2 = 2^2$$
$$x^2 + 2 = 4$$
$$x^2 = 2$$
$$x = \sqrt{2}$$

Then

$$b_2 = 5 + \sqrt{2} + \sqrt{2}$$
$$= 5 + 2\sqrt{2}$$

Finally, we use the formula for the area of a trapezoid.

$$A = \frac{1}{2}h(b_1 + b_2)$$

$$= \frac{1}{2}\sqrt{2}(5 + 5 + 2\sqrt{2})$$

$$= \frac{1}{2}\sqrt{2}(10 + 2\sqrt{2})$$

$$= (5\sqrt{2} + 2) \text{ m}^2$$

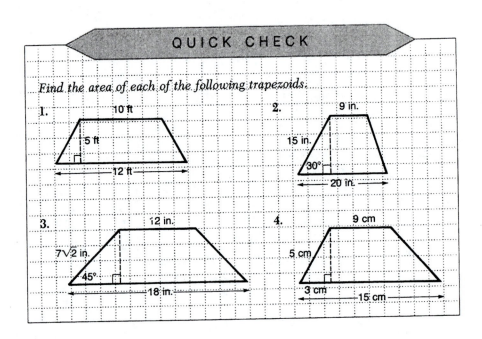

QUICK CHECK

Find the area of each of the following trapezoids.

1. 10 ft, 5 ft, 12 ft

2. 9 in., 15 in., 30°, 20 in.

3. 12 in., $7\sqrt{2}$ in., 45°, 18 in.

4. 9 cm, 5 cm, 3 cm, 15 cm

---

5 ▶ AREA OF A REGULAR HEXAGON

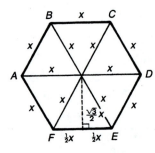

FIGURE 5.12

The area of a regular hexagon can be obtained by sectioning the hexagon into six equilateral triangles and then multiplying the area of one equilateral triangle by 6. In Figure 5.12 regular hexagon *ABCDEF* is divided into six equilateral triangles. Its area is six times the area of one equilateral triangle. Using the formula for an equilateral triangle,

$$\text{Area of regular hexagon } ABCDEF = 6\left(\frac{\sqrt{3}x^2}{4}\right)$$

$$= \frac{3\sqrt{3}x^2}{2}$$

ANSWERS

1. 55 sq ft   **2.** 108.75 sq in.   **3.** 108.75 sq in.   **4.** 48 sq cm

AREA OF A REGULAR HEXAGON

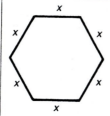

$$A = \frac{3\sqrt{3}}{2}x^2$$

EXAMPLE 16   Find the area of a regular hexagon whose sides are each 26 yards.

Solution

$$A = \frac{3\sqrt{3}}{2}x^2$$

$$= \frac{3\sqrt{3}}{2}(26)^2$$

$$= \frac{3\sqrt{3}}{2}(676)$$

$$= 1014\sqrt{3} \text{ yd}^2$$

◄

QUICK CHECK

1. Find the area of a regular hexagon whose sides each measure 4 inches.

2. The distance from the center of a regular hexagon to one of its vertices is 9 inches. What is the area?

3. The distance from the center of a regular hexagon to one of its sides is $5\sqrt{3}$ inches. Find the area.

6 ▶ AREA OF A CIRCLE

As in the formula for the circumference of a circle, the formula for the area of a circle involves the irrational number π. In this book we use 3.14 as the approximate value of π.

ANSWERS

1. $24\sqrt{3}$ sq in.   2. $(121.5)\sqrt{3}$ sq in.   3. $150\sqrt{3}$ sq in.

AREA OF A CIRCLE

$A = \pi r^2$

EXAMPLE 17   Using $\pi \approx 3.14$, find the area of a circle whose diameter is 6 meters.

Solution
$$A = \pi r^2$$
$$\approx (3.14)(3)^2$$
$$\approx (3.14)(9)$$
$$\approx 28.26 \text{ m}^2$$ ◀

EXAMPLE 18   Using $\pi = 3.14$, the area of a circle was calculated as 21.98 square centimeters. What is the length of the radius?

Solution
$$A = \pi r^2$$
$$21.98 = 3.14r^2$$
$$7 \approx r^2$$
$$r \approx \sqrt{7} \text{ cm}$$ ◀

QUICK CHECK

1. A circle has a radius of 12 inches. Find the approximate area (rounded to the hundredths place).

2. Find the exact area of a circle whose diameter is 12 inches.

3. The circumference of a circle is 3π feet. Find the exact area.

In this section we have left our answers in terms of square root radicals. You may, however, use your calculator to find decimal approximations for these square roots and then leave your answers in decimal form. Remember that your calculator has a $\boxed{\sqrt{x}}$ button on it. If you want to find $\sqrt{3}$, enter 3 and press the $\boxed{\sqrt{x}}$ button. The number 1.7320508 will be displayed, which is a seven-decimal approximation for $\sqrt{3}$. Remember that $\sqrt{3}$ is irrational and in decimal form does not repeat and does not terminate. As a general rule of thumb, you can round your answers to two decimal places.

ANSWERS

1. 452.16 sq in.   2. 36π sq in.   3. $\frac{9}{4}\pi$ sq ft

A

For Exercises 1–39, find the area.

1.

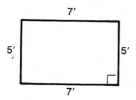

1. _____

2.

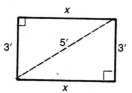

2. _____

3.

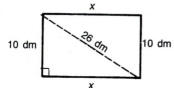

3. _____

4.

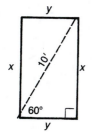

4. _____

5.

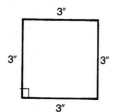

5. _____

6.

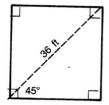

6. _____

7.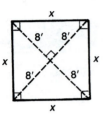

8.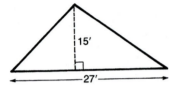

9.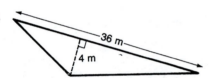

10.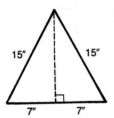

11.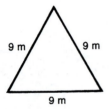

12.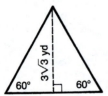

13.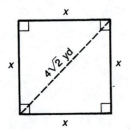

14.

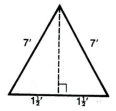

14. ——————————

15.

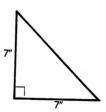

15. ——————————

16.

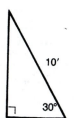

16. ——————————

17.

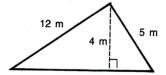

17. ——————————

18.

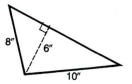

18. ——————————

19.

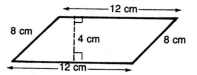

19. ——————————

20.

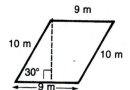

20. ——————————

21.

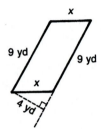

21. _____

22.

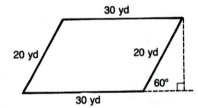

22. _____

23.

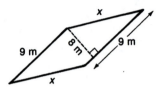

23. _____

24.

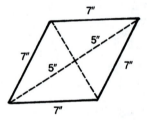

24. _____

25.

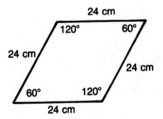

25. _____

26.

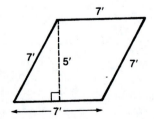

26. _____

27.

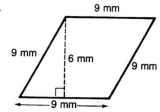

28.

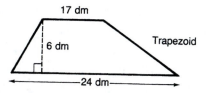

Trapezoid

29.

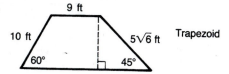

Trapezoid

30.

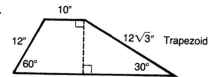

Trapezoid

31.

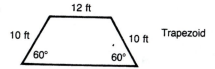

Trapezoid

32.

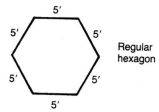

Regular
hexagon

33.

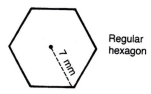

Regular
hexagon

34.

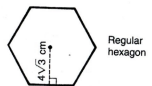

Regular
hexagon

35.

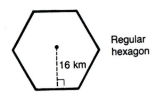

Regular hexagon

16 km

35. _____

36.

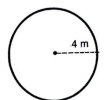

.4 m

36. _____

37.

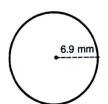

6.9 mm

37. _____

38.

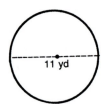

11 yd

38. _____

39.

26.47 m

39. _____

40. If the area of a rectangle is 15 square meters and the width is 5 meters, find the length.

40. _____

41. The area of a square is 25 square meters. Find the length of its sides.

41. _____

42. The area of a rectangle is 1.76 square centimeters and the length is 2.1 centimeters. Find the width.

42. _____

43. The area of a square is 1.44 square millimeters. Find the length of its sides.

43. _____

44. The area of an equilateral triangle is $25\sqrt{3}$. Find the length of its sides.

44. _____

Perform the following unit conversions.

45. 26 in.^{2} = _____ ft^2

45. _____

46. 350 in.^{2} = _____ ft^2

46. _____

47. 4 yd^2 = _____ ft^2

47. _____

48. $16 \text{ yd}^2 = \underline{\hspace{1.5cm}} \text{ ft}^2$

49. $32 \text{ ft}^2 = \underline{\hspace{1.5cm}} \text{ yd}^2$

50. $75 \text{ ft}^2 = \underline{\hspace{1.5cm}} \text{ yd}^2$

51. $40 \text{ ft}^2 = \underline{\hspace{1.5cm}} \text{ in.}^2$

52. $16 \text{ ft}^2 = \underline{\hspace{1.5cm}} \text{ in.}^2$

53. $1 \text{ mi}^2 = \underline{\hspace{1.5cm}} \text{ yd}^2$

54. $1.5 \text{ mi}^2 = \underline{\hspace{1.5cm}} \text{ yd}^2$

55. $25 \text{ m}^2 = \underline{\hspace{1.5cm}} \text{ dm}^2$

56. $15 \text{ m}^2 = \underline{\hspace{1.5cm}} \text{ dm}^2$

57. $36 \text{ hm}^2 = \underline{\hspace{1.5cm}} \text{ m}^2$

58. $50 \text{ hm}^2 = \underline{\hspace{1.5cm}} \text{ m}^2$

59. $2000 \text{ mm}^2 = \underline{\hspace{1.5cm}} \text{ cm}^2$

60. $5000 \text{ mm}^2 = \underline{\hspace{1.5cm}} \text{ cm}^2$

61. $300 \text{ cm}^2 = \underline{\hspace{1.5cm}} \text{ dam}^2$

62. $500 \text{ cm}^2 = \underline{\hspace{1.5cm}} \text{ dam}^2$

63. $200 \text{ dm}^2 = \underline{\hspace{1.5cm}} \text{ mm}^2$

64. $600 \text{ dm}^2 = \underline{\hspace{1.5cm}} \text{ mm}^2$

65. $12 \text{ yd}^2 = \underline{\hspace{1.5cm}} \text{ m}^2$

66. $26 \text{ yd}^2 = \underline{\hspace{1.5cm}} \text{ m}^2$

67. $73 \text{ ft}^2 = \underline{\hspace{1.5cm}} \text{ dm}^2$

68. $96 \text{ ft}^2 = \underline{\hspace{1.5cm}} \text{ dm}^2$

69. $2.3 \text{ in.}^2 = \underline{\hspace{1.5cm}} \text{ cm}^2$

70. $7.8 \text{ in.}^2 = \underline{\hspace{1.5cm}} \text{ cm}^2$

71. $2.9 \text{ km}^2 = \underline{\hspace{1.5cm}} \text{ mi}^2$

72. $6.3 \text{ km}^2 = \underline{\hspace{1.5cm}} \text{ mi}^2$

73. $46 \text{ dam}^2 = \underline{\hspace{1.5cm}} \text{ ft}^2$

74. $79 \text{ dam}^2 = \underline{\hspace{1.5cm}} \text{ ft}^2$

75. $15 \text{ ac} = \underline{\hspace{1.5cm}} \text{ yd}^2$

76. $23 \text{ ac} = \underline{\hspace{1.5cm}} \text{ yd}^2$

77. $6 \text{ ha} = \underline{\hspace{1.5cm}} \text{ ac}$

78. $9 \text{ ha} = \underline{\hspace{1.5cm}} \text{ ac}$

79. $22 \text{ ha} = \underline{\hspace{1.5cm}} \text{ ac}$

80. $17 \text{ ha} = \underline{\hspace{1.5cm}} \text{ ac}$

81. $3 \text{ ac} = \underline{\hspace{1.5cm}} \text{ ha}$

82. $5 \text{ ac} = \underline{\hspace{1.5cm}} \text{ ha}$

83. A rectangle has a length of 4 inches and a width of 9 inches. Find the area in square feet.

48. _____

49. _____

50. _____

51. _____

52. _____

53. _____

54. _____

55. _____

56. _____

57. _____

58. _____

59. _____

60. _____

61. _____

62. _____

63. _____

64. _____

65. _____

66. _____

67. _____

68. _____

69. _____

70. _____

71. _____

72. _____

73. _____

74. _____

75. _____

76. _____

77. _____

78. _____

79. _____

80. _____

81. _____

82. _____

83. _____

84. A rectangle has a length of 3 feet and a width of 17 feet. Find the area in square yards.

84. _____

85. The sides of a square measure 3 decimeters. Find the area in square meters.

85. _____

86. The sides of a square measure 3 meters. Find the area in square centimeters.

86. _____

87. The base of a triangle is 15 inches and its height is 26 inches. Find the area in square feet.

87. _____

88. The base of a triangle is 19 feet and its altitude is 14 feet. Find the area in square inches.

88. _____

89. A parallelogram has a base of 15 feet and a height of 8 feet. Find the area in square inches.

89. _____

90. A parallelogram has a base of 12 yards and a height of 4 yards. Find the area in square meters.

90. _____

91. The side of a rhombus is 13 centimeters and the height is $9\frac{3}{13}$ centimeters. Find the area in square inches.

91. _____

92. The side of a rhombus is 5 inches and the height is $4\frac{4}{5}$ inches. Find the area in square centimeters.

92. _____

93. A tract of land contains 25 acres. What is the area in hectares?

93. _____

94. A rectangular tract of land contains 260 acres. What is the area in hectares?

94. _____

95. A rectangular tract of land measures 200 yards by 250 yards. What is the area in acres?

95. _____

96. A tract of land measures 400 yards by 350 yards. What is the area in acres?

96. _____

97. How many acres are in 3 sections of land?

97. _____

98. How many acres are in 5 sections of land?

98. _____

99. The Gobi Desert has an area of 400,000 square miles. How many sections of land does it cover?

99. _____

100. Australia covers 2,941,500 square miles. How many sections of land does Australia contain?

100. _____

B

101. A rectangular room measures 12 feet 3 inches by 15 feet. How many square yards of carpet are needed for this room? If carpet costs $13 per square yard, how much will it cost to carpet the room?

101. _____

102. A rectangular room measures 9 feet 6 inches by 10 feet 4 inches. How many square feet are in this room? If the floor is to be covered with tiles that are 8 inches by 8 inches, how many tiles will be needed? If the tiles cost $0.37 each, how much will it cost to tile the floor?

102. _____

103. An acre is 4840 square yards. A farmer has 96 acres and plans to plant it in grass by scattering the seeds on the surface. The seeds cost $1.20 per pound and 1 pound of seeds is enough to plant 125 square yards. How many pounds should he buy?

103. _____

104.   Find the area of the figure shown.   104. _____

105.   Find the area of the figure shown.   105. _____

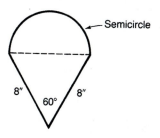

106.   Find the area of the shaded region shown if $r = 8$ dm and   106. _____
       $R = 15$ dm.

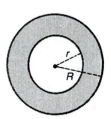

107.   Find the area of the shaded region if each circle has a radius of 1 inch.   107. _____
       (Hint: Draw an equilateral triangle that has its vertices at the centers
       of the three corner circles.)

5.2 SURFACE AREA

OBJECTIVES

▶ **1** Find the surface area of a sphere and a cone.

▶ **2** Find the surface area of a prism and a pyramid.

▶ **3** Find the surface area of a cylinder.

▶ **1** CONES AND SPHERES

The term **surface area** refers to the amount of area on the surface of a three-dimensional space figure such as a pyramid, prism, or sphere.

Total surface area refers to the area of all surfaces of a space figure, including its bases. **Lateral surface area** refers to the total surface area minus the area of the base or bases.

The formulas for the surface area of a cone and a sphere involve the irrational constant π.

SURFACE AREA

Sphere

Surface area $= 4\pi r^2$

where r is the radius of the sphere.

Cone

Lateral surface area $= \pi rs$

where r is the radius of the base and s is the slant height.

Total surface area $= \pi rs + \pi r^2$

$$= \pi r(s + r)$$

EXAMPLE 19 Find the surface area of a sphere whose radius is 3.7 cm. Use $\pi \approx 3.14$ and round your answer to the nearest hundredth.

Solution
$$\text{Surface area} = 4\pi r^2$$
$$\approx 4(3.14)(3.7)^2$$
$$\approx 171.95 \text{ cm}^2$$

EXAMPLE 20 Find the lateral surface area of a cone whose base has a radius of 3 inches and whose slant height is 5 inches. Use $\pi \approx 3.14$.

Solution
$$\text{Lateral surface area} = \pi r s$$
$$\approx (3.14)(3)(5)$$
$$\approx 47.1 \text{ in.}^2$$

EXAMPLE 21 Find the total surface area of the cone in the figure shown. Use $\pi \approx 3.14$.

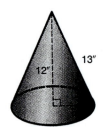

Solution
$$\text{Total surface area} = \pi r s + \pi r^2$$
$$\approx (3.14)(5)(13) + (3.14)(5)^2$$
$$\approx 204.1 + 78.5$$
$$\approx 282.6 \text{ in.}^2$$

QUICK CHECK

1. A sphere has a radius of 14 inches. What is the exact surface area?

2. Find the exact surface area of a sphere whose diameter is 23 feet.

3. A cone has a slant height of 15 inches and a radius of 12 inches. What is the exact lateral area?

4. A cone has a height of 12 inches and a base radius of 9 inches. What is the approximate total surface area?

ANSWERS

1. 784π in.^{2} 2. 529π ft^2 3. 180π in.^{2} 4. 678.24 in.^{2}

▶ PRISMS AND PYRAMIDS

The surface area of a prism and a pyramid may be determined by using the concepts developed in the preceding section to determine the area of the faces and bases. Once these areas are established, the surface area is obtained by adding the areas.

_____ E X A M P L E   2 2   Find the total surface area of the right figure shown. Both bases are regular hexagons.

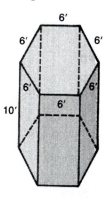

Solution   First we find the area of the bases.

$$\text{Area of equilateral triangle} = \frac{1}{2}(6)(3\sqrt{3})$$
$$= 9\sqrt{3}$$

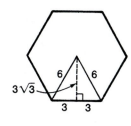

$$\text{Area of hexagon} = 6(9\sqrt{3})$$
$$= 54\sqrt{3}$$
$$\text{Area of bases} = 2(54\sqrt{3})$$
$$= 108\sqrt{3}$$

Next we find the area of the lateral faces. Since each of the six lateral faces is a rectangle, their total area is

$$\text{Area of lateral faces} = 6(6 \cdot 10)$$
$$= 6(60)$$
$$= 360$$

$$\text{Total surface area} = 108\sqrt{3} + 360$$
$$\approx 547.06 \text{ ft}^2$$

◀

_____ E X A M P L E   2 3   Find the total surface area of the right pyramid shown.

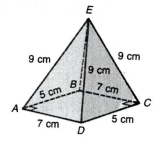

Solution   First find the lateral area by determining the area of each triangular face.

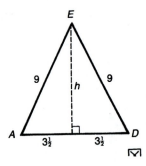

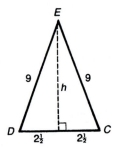

$$\left(3\frac{1}{2}\right)^2 + h^2 = 9^2 \qquad\qquad h^2 + \left(2\frac{1}{2}\right)^2 = 9^2$$

$$h^2 = 81 - \frac{49}{4} \qquad\qquad h^2 = 81 - \frac{25}{4}$$

$$= \frac{275}{4} \qquad\qquad h^2 = \frac{299}{4}$$

$$h = \sqrt{\frac{25 \cdot 11}{4}} \qquad\qquad h = \frac{\sqrt{299}}{2}$$

$$= \frac{5\sqrt{11}}{2}$$

$$\text{Area of } \Delta ADE = \frac{1}{2}(7)\left(\frac{5\sqrt{11}}{2}\right) \qquad \text{Area of } \Delta DEC = \frac{1}{2}(5)\left(\frac{\sqrt{299}}{2}\right)$$

$$= \frac{35\sqrt{11}}{4} \qquad\qquad = \frac{5(\sqrt{299})}{4}$$

$$\text{Lateral area of pyramid} = 2\left(\frac{35\sqrt{11}}{4}\right) + 2\left(\frac{5\sqrt{299}}{4}\right)$$

$$= \frac{35\sqrt{11}}{2} + \frac{5\sqrt{299}}{2}$$

$$\approx 101.27 \text{ cm}^2$$

Now find the area of the base of the pyramid.

$$\text{Area of base} = 7 \cdot 5$$

$$= 35$$

$$\text{Total surface area} \approx 101.27 + 35$$

$$\approx 136.27 \text{ cm}^2$$

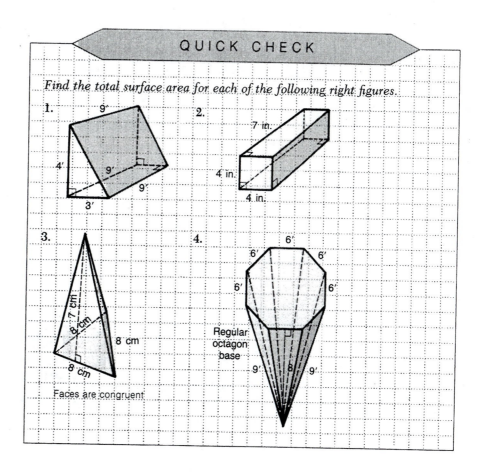

QUICK CHECK

Find the total surface area for each of the following right figures.

1.

2.

3.

Faces are congruent

4.

Regular octagon base

---

③ ▶ CYLINDERS

The total surface area of a right circular cylinder can be determined by adding the area of its bases to the area of its side surface. Since its bases are circles and the area of a circle is πr^2, the area of its two bases is $2\pi r^2$. The area of its side surface is the same as the area of a rectangle of the same height as the cylinder and of a length equal to the circumference of the cylinder. See Figure 5.13.

FIGURE 5.13
Right circular cylinder:
Total surface area $= 2\pi r^2 + 2\pi rh$

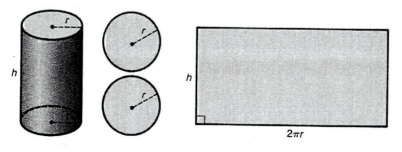

ANSWERS

1. $120\,\text{ft}^2$ **2.** $144\,\text{in.}^2$ **3.** $16\sqrt{3} + 84\,\text{cm}^2$ **4.** $144\sqrt{2} + 54\sqrt{3}\,\text{ft}^2$

EXAMPLE 24   Find the total surface area of a right circular cylinder whose base has a radius of 4 meters and whose height is $\frac{1}{2}$ meter.

Solution   Total surface area $= 2\pi r^2 + 2\pi rh$

$$\approx 2(3.14)(16) + 2(3.14)(4)\left(\frac{1}{2}\right)$$

$$\approx 100.48 + 12.56$$

$$\approx 113.04 \text{ m}^2$$

QUICK CHECK

Find the exact total surface area for each of the following right circular cylinders.

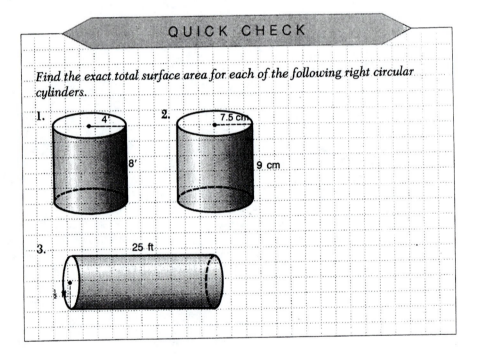

1.   4'   8'

2.   7.5 cm   9 cm

3.   25 ft $\frac{1}{2}'$

ANSWERS
1. 96π ft² 　2. 247.5π cm² 　3. 25.5π ft²

A

Find the total surface area for each of the following. Assume all figures are right.

1.

1. _____

2.

2. _____

3.

3. _____

4.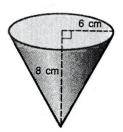

4. _____

5.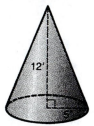

5. _____

6.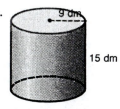

6. _____

7.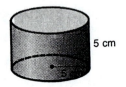

5 cm

7. _____

8.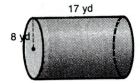

17 yd

8 yd

8. _____

9.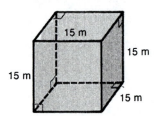

15 m

15 m

15 m

15 m

9. _____

10.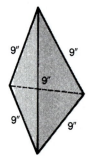

9″   9″

9″

9″   9″

10. _____

11.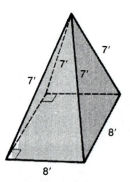

7′

7′   7′

7′

8′

8′

11. _____

12.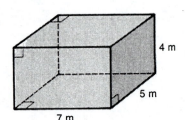

4 m

5 m

7 m

12. _____

13.

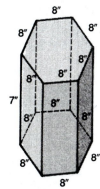

Bases are regular hexagons

13. _____

14.

14. _____

15.

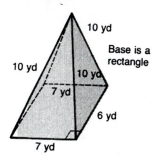

Base is a rectangle

15. _____

16.

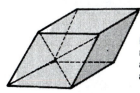

Each base and face is a rhombus of side 10 inches and diagonals 12 inches and 16 inches

16. _____

17.

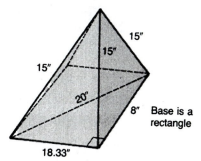

Base is a rectangle

17. _____

18.

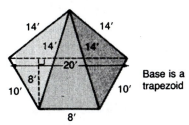

Base is a trapezoid

18. _____

19.

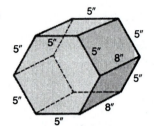

Bases are
regular
hexagons

19. _____

20.

20. _____

21.

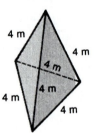

21. _____

22.

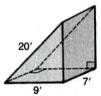

22. _____

23.

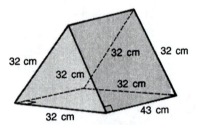

23. _____

24.

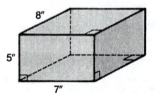

24. _____

25.

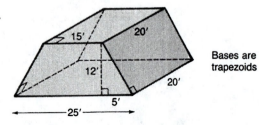

Bases are
trapezoids

25. _____

<u>B</u>

26. A room with 8-foot ceilings measures 14 feet by 18 feet. It has three windows that measure 3 feet by 5 feet and two doors that measure 3 feet by 6 feet 8 inches. What is the total surface area of the walls and ceiling?

26. _____

27. A gallon of paint will cover 288 square feet of surface (for one coat of paint). How much paint will it take to paint the walls and ceiling in problem 26 if two coats are applied?

27. _____

28. A pup tent is to be constructed from canvas with the dimensions as shown in the figure. Without allowing for seams, how much canvas is required? (The tent does not have a floor.)

28. _____

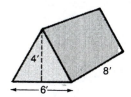

29. A city has a cylindrical storage tank for water that measures 45 feet across and is 36 feet tall. The city plans to paint the sides and top. What total surface area are they planning to paint?

29. _____

5.3 VOLUME OF SPACE FIGURES

OBJECTIVE

▶ **1** Find the volume of a prism, cylinder, pyramid, cone, and sphere; convert cubic units.

▶ **1** VOLUME

Volume is the amount of space contained within a three-dimensional space figure and is measured in cubic units. A cubic unit such as a cubic inch is the size of a cube that measures 1 inch on a side. The relative size of a cubic inch and a cubic centimeter is shown in Figure 5.14.

FIGURE 5.14
Cubic inch and cubic centimeter

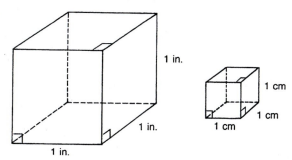

To find the volume of the prism means to find the number of cubic units it contains. Figure 5.15 shows a rectangular prism. Counting the cubic centimeters in the prism reveals that it contains 40 cubic centimeters; thus, the prism's volume is 40 cubic centimeters. The volume can also be determined by multiplying its length, width, and height; that is, $5 \cdot 4 \cdot 2 = 40$ cubic centimeters.

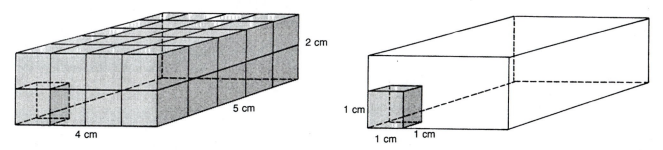

FIGURE 5.15
Rectangular prism: volume of the rectangular prism is 40 cubic centimeters

Volume is measured in cubic units. Cubic units are abbreviated by preceding the unit name with "cu" or by using the exponent 3 on the unit. Thus, 45 cu ft and 45 ft³ are both read "45 cubic feet."

Volume can be measured using the cube of any metric or American unit of measure for length. Unit conversions involving volume are performed in the same way we did for area, except that the units are cubed instead of squared.

<u>EXAMPLE 25</u>   Convert 0.008 cubic feet to cubic inches.

Solution $0.008 \text{ ft}^3 = 0.008 \text{ ft}^3 \left(\dfrac{12 \text{ in.}}{1 \text{ ft}}\right)^3$

$= 0.008 \text{ ft}^3 \cdot \dfrac{1728 \text{ in.}^3}{1 \text{ ft}^3}$

$= 13.82 \text{ in.}^3$ ◀

<u>EXAMPLE 26</u>   Perform each of the following unit conversions.

1. $15 \text{ m}^3 = $ _____ cm^3 2. $25 \text{ yd}^3 = $ _____ ft^3
3. $360 \text{ in.}^3 = $ _____ dm^3

Solution   1. $15 \text{ m}^3 = 15 \text{ m}^3 \left(\dfrac{100 \text{ cm}}{1 \text{ m}}\right)^3$ 2. $25 \text{ yd}^3 = 25 \text{ yd}^3 \left(\dfrac{3 \text{ ft}}{1 \text{ yd}}\right)^3$

$= 15 \text{ m}^3 \cdot \dfrac{1,000,000 \text{ cm}^3}{1 \text{ m}^3}$ $= 25 \text{ yd}^3 \cdot \dfrac{27 \text{ ft}^3}{1 \text{ yd}^3}$

$= 15,000,000 \text{ cm}^3$ $= 675 \text{ ft}^3$

3. $360 \text{ in.}^3 = 360 \text{ in.}^3 \left(\dfrac{2.54 \text{ cm}}{1 \text{ in.}}\right)^3 \left(\dfrac{1 \text{ dm}}{10 \text{ cm}}\right)^3$

$= 360 \text{ in.}^3 \cdot \dfrac{16.39 \text{ cm}^3}{1 \text{ in.}^3} \cdot \dfrac{1 \text{ dm}^3}{1000 \text{ cm}^3}$

$= 5.9 \text{ dm}^3$ ◀

The formulas for finding the volume of right prisms, cylinders, pyramids, cones, and spheres are detailed below. Notice that we use the formulas from Section 5.1 to find the area of the bases and that we multiply these areas by the heights of the space figures. Notice also that the volume of space figures that come to a point (cones and pyramids) have an additional factor of $\frac{1}{3}$.

VOLUME

Volume of a Right Prism or Cylinder

$V = A \cdot h$

where A represents the area of the base and h represents the length of the lateral edge or height.

Volume of a Right Pyramid or Cone

$V = \dfrac{1}{3} A \cdot h$

where A represents the area of the base and h represents the height, which is the perpendicular distance from the vertex to the base.

Volume of a Sphere

$V = \dfrac{4}{3} \pi r^3$

where r is the radius.

EXAMPLE 27   Find the volume of the sphere shown. Use $\pi \approx 3.14$ and round your answer to the nearest hundredth.

Solution
$$V = \left(\frac{4}{3}\right)\pi r^3$$

$$\approx \left(\frac{4}{3}\right)(3.14)(7^3)$$

$$\approx \left(\frac{4}{3}\right)(3.14)(343)$$

$$\approx 1436.03 \text{ cm}^3$$

EXAMPLE 28   Find the volume of the figure shown.

Solution
$$V = \left(\frac{1}{3}\right)A \cdot h$$

$$= \left(\frac{1}{3}\right)(\pi r^2)h$$

$$\approx \left(\frac{1}{3}\right)(3.14)(4^2)(7)$$

$$\approx \left(\frac{1}{3}\right)(3.14)(16)(7)$$

$$\approx 117.23 \text{ in.}^3$$

EXAMPLE 29   Find the volume of the right figure shown. Round your answer to the nearest hundredths place.

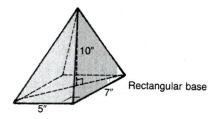

Rectangular base

Solution $V = \left(\dfrac{1}{3}\right)(l \cdot w)(h)$

$= \left(\dfrac{1}{3}\right)(5 \cdot 7)(10)$

$= 116.67 \text{ in.}^3$ ◀

EXAMPLE 30   Find the volume of the figure shown and round your answer to the nearest hundredths place. The bases are regular hexagons.

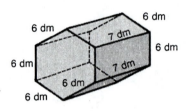

Solution $A = 6\left[\left(\dfrac{1}{2}\right)(6)(3\sqrt{3})\right]$

$= 54\sqrt{3}$

$V = A \cdot h$

$= 54\sqrt{3} \cdot 7$

$\approx 654.72 \text{ dm}^3$ ◀

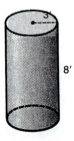

EXAMPLE 31   Find the volume of the figure shown. Use $\pi \approx 3.14$.

Solution $V = A \cdot h$

$= \pi r^2 h$

$\approx (3.14)(3^2)(8)$

$\approx (3.14)(9)(8)$

$\approx 226.08 \text{ ft}^3$ ◀

```
                ┌─────────────────────────────┐
                │        QUICK CHECK          │
                └─────────────────────────────┘
```

Perform the following unit conversions.

1. $315 \text{ in.}^3 =$ _____ ft^3 2. $3 \text{ m}^3 =$ _____ dam^3

3. $7 \text{ ft}^3 =$ _____ m^3

4. If a sphere has a radius of 7 meters, find its approximate volume.

5. Find the approximate volume of a cone that has a slant height of 7 inches and a base radius of 6 inches.

6. The area of the base of a rectangular parallelepiped is 23 square feet and the height is 5 feet. Find the volume.

7. Find the exact volume of a cylinder that has a diameter of 4 inches and a height of 8 inches.

8. A right prism has equilateral triangular bases 3 inches on a side and a height of 5 inches. Find the approximate volume.

9. A right pyramid has a square base that is 4 meters on a side and the height is 5 meters. Find the volume.

ANSWERS

1. 0.1823 ft^3 2. 0.003 dam^3 3. 0.198 m^3 4. 1436.03 m^3 5. 135.86 in.^3 6. 115 ft^3
7. $32\pi \text{ in.}^3$ 8. 19.49 in.^3 9. $26\frac{2}{3} \text{ m}^3$

A

Perform the indicated unit conversions.

ANSWERS

1. $39 \text{ in.}^3 = \underline{\hspace{1.5cm}} \text{ ft}^3$

2. $46 \text{ in.}^3 = \underline{\hspace{1.5cm}} \text{ ft}^3$

3. $900 \text{ ft}^3 = \underline{\hspace{1.5cm}} \text{ yd}^3$

4. $2176 \text{ ft}^3 = \underline{\hspace{1.5cm}} \text{ yd}^3$

5. $463 \text{ yd}^3 = \underline{\hspace{1.5cm}} \text{ ft}^3$

6. $965 \text{ yd}^3 = \underline{\hspace{1.5cm}} \text{ ft}^3$

7. $263 \text{ ft}^3 = \underline{\hspace{1.5cm}} \text{ in.}^3$

8. $765 \text{ ft}^3 = \underline{\hspace{1.5cm}} \text{ in.}^3$

9. $92 \text{ m}^3 = \underline{\hspace{1.5cm}} \text{ hm}^3$

10. $63 \text{ m}^3 = \underline{\hspace{1.5cm}} \text{ hm}^3$

11. $200 \text{ cm}^3 = \underline{\hspace{1.5cm}} \text{ dm}^3$

12. $743 \text{ cm}^3 = \underline{\hspace{1.5cm}} \text{ dm}^3$

13. $525 \text{ dm}^3 = \underline{\hspace{1.5cm}} \text{ hm}^3$

14. $968 \text{ dm}^3 = \underline{\hspace{1.5cm}} \text{ hm}^3$

15. $2000 \text{ mm}^3 = \underline{\hspace{1.5cm}} \text{ cm}^3$

16. $6000 \text{ mm}^3 = \underline{\hspace{1.5cm}} \text{ cm}^3$

17. $50 \text{ in.}^3 = \underline{\hspace{1.5cm}} \text{ cm}^3$

18. $96 \text{ in.}^3 = \underline{\hspace{1.5cm}} \text{ dm}^3$

19. $23 \text{ ft}^3 = \underline{\hspace{1.5cm}} \text{ dm}^3$

20. $95 \text{ ft}^3 = \underline{\hspace{1.5cm}} \text{ dm}^3$

21. $42 \text{ m}^3 = \underline{\hspace{1.5cm}} \text{ yd}^3$

22. $78 \text{ m}^3 = \underline{\hspace{1.5cm}} \text{ yd}^3$

23. $5 \text{ km}^3 = \underline{\hspace{1.5cm}} \text{ mi}^3$

24. $6 \text{ km}^3 = \underline{\hspace{1.5cm}} \text{ mi}^3$

1. _____

2. _____

3. _____

4. _____

5. _____

6. _____

7. _____

8. _____

9. _____

10. _____

11. _____

12. _____

13. _____

14. _____

15. _____

16. _____

17. _____

18. _____

19. _____

20. _____

21. _____

22. _____

23. _____

24. _____

For Exercises 25–49 find the exact volume of each right figure.

25.

4 m

25. _____

26.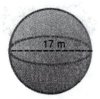

26. _____

27.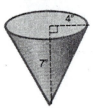

27. _____

28.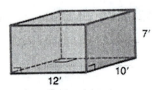

28. _____

29.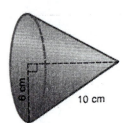

29. _____

30.

30. _____

31.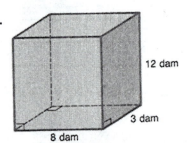

31. _____

32.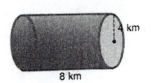

32. _____

33.

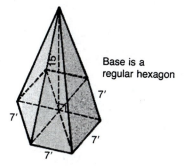

Base is a
regular hexagon

15

7' 7' 7' 7' 7' x

33. _____

34.

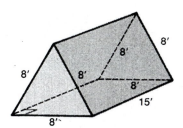

8' 8' 8' 8' 8' 8' 15'

34. _____

35.

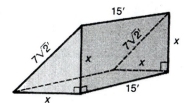

15' $7\sqrt{2}$ $7\sqrt{2}$ x x x 15' x

35. _____

36.

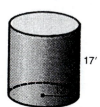

17'

36. _____

37.

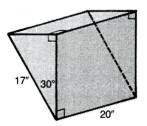

17" 30° 20"

37. _____

38.

30° 10"

38. _____

39.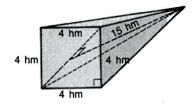

40.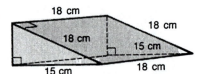

41.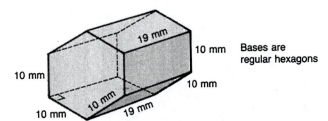

Bases are
regular hexagons

42.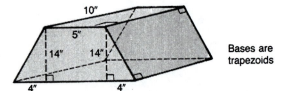

Bases are
trapezoids

43.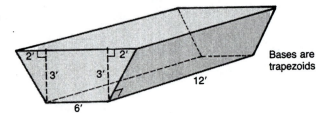

Bases are
trapezoids

44.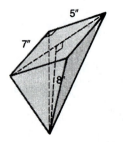

45.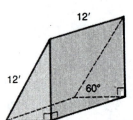

46.

47.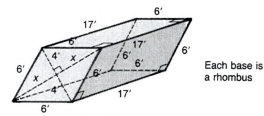

Each base is a rhombus

47. _____

48.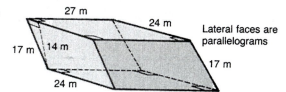

Lateral faces are parallelograms

48. _____

49.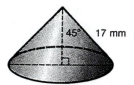

49. _____

B

50. Find the volume of the figure shown.

50. _____

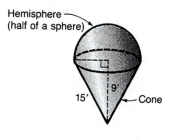

Hemisphere (half of a sphere)

Cone

51. Find the volume of the figure shown.

51. _____

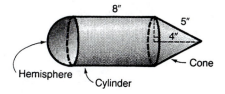

Hemisphere   Cylinder   Cone

52. A tennis ball has a 1-inch radius. Find the volume of the smallest cylindrical can required to hold four tennis balls.

52. _____

53. What is the volume of the air in the can of tennis balls in Exercise 52?

53. _____

ANSWERS

1. Find the area of a square whose diagonal measures 15 centimeters.   1. _____

2. Find the area of a rhombus whose diagonals measure 14 meters and 16 meters, respectively.   2. _____

3. The distance between the parallel sides of a trapezoid is 4 inches. The bases measure 3 inches and 7 inches. Find the area.   3. _____

4. One pair of opposite sides of a parallelogram each measure 15 hectometers. The distance between these parallel sides is 12 hectometers. Find the area of the parallelogram.   4. _____

5. One side of a triangle measures 15 yd and the altitude drawn to this side measures 4 feet. Find the area of the triangle.   5. _____

6. One side of a rectangle measures 36.4 inches and the diagonal measures 52.6 inches. What is the area?   6. _____

7. The distance from the center of a regular hexagon to one of its sides is 4 centimeters. Find the area.   7. _____

8. Find the area of a circle whose circumference is 6.28 meters.   8. _____

9. The radius of the base of a cone is 15 decimeters and the height is 120 centimeters. Find the lateral surface area.   9. _____

10. The surface area of a sphere is 265 square centimeters. Find the radius.   10. _____

11. Find the total surface area of a cylinder whose height is 7 feet and whose diameter is 21 feet.   11. _____

12. Find the lateral surface area of a right pyramid whose base is a square 15 decimeters on a side and whose lateral edges are 19 decimeters.   12. _____

13. Find the total surface area of a rectangular prism that has a length of 15 meters, a width of 14 meters, and a height of 4 meters.   13. _____

14. Find the volume of a sphere whose diameter is 36 yards.   14. _____

15. Find the volume of a cone whose slant height is 26 centimeters and whose base has a 20-centimeter diameter.   15. _____

16. If the diameter of a cylinder is 0.17 hectometers and the height is 4 meters, find the volume.   16. _____

17. A right hexagonal prism has a height of 14 inches and each side of the bases measures 2 inches. Find the volume.   17. _____

18. Find the volume of a cube that measures 5 centimeters on each edge.   18. _____

19. A room measures 20 feet by 18 feet by 9 feet. Find the volume.   19. _____

Perform the following unit conversions.

20. $4 \text{ in.}^2 = $ _____ ft^2 20. _____

21. $25 \text{ ft}^2 = $ _____ yd^2 21. _____

22. $300 \text{ cm}^2 = $ _____ m^2 22. _____

23. $32 \text{ ft}^2 = $ _____ m^2 23. _____

24. $35 \text{ ft}^3 = $ _____ yd^3 24. _____

25. $64 \text{ m}^3 = $ _____ in.^3 25. _____

ANSWERS

1. Find the area of a circle whose radius is 8 feet.

2. Find the area of a circle whose diameter is 4 centimeters.

Find the area for each of the following figures.

3.

4 ft
3 ft
4 ft
3 ft

4.

4 in
30°

5.

27 ft
9 ft
Parallelogram

6.

8 in.   12 in.
12 in.   8 in.

7.

14 mm
Trapezoid
6 mm
18 mm

Find the volume for each of the following figures.

8.

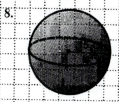

1. _____

2. _____

3. _____

4. _____

5. _____

6. _____

7. _____

8. _____

243

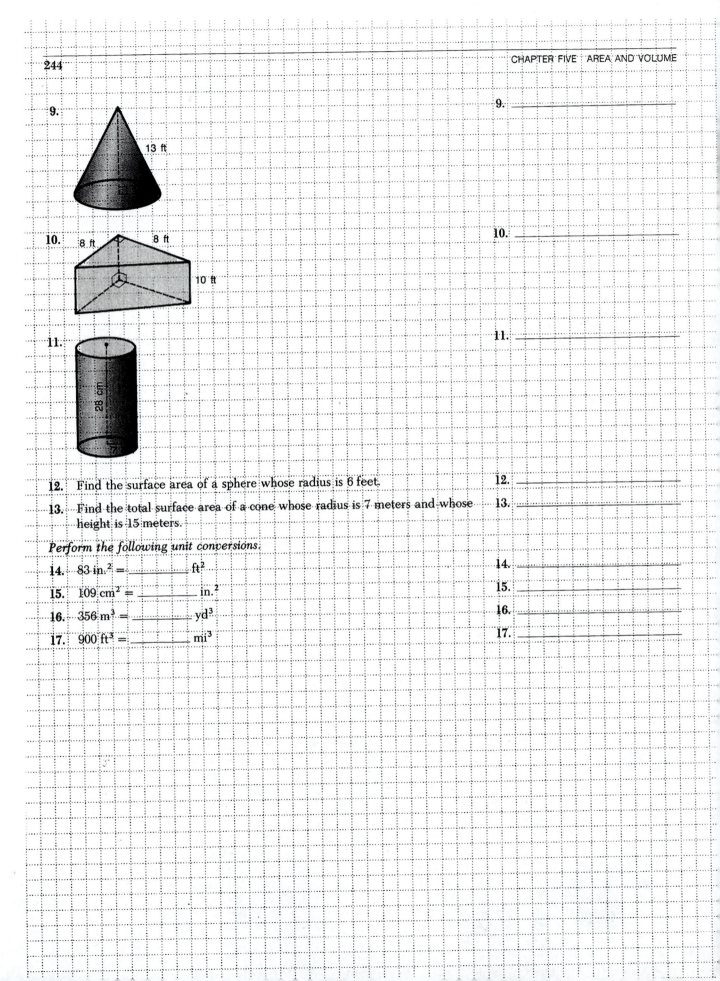

9.

13 ft

9. _____

10. 8 ft   8 ft

10 ft

10. _____

11.

28 cm

11. _____

12. Find the surface area of a sphere whose radius is 6 feet.

12. _____

13. Find the total surface area of a cone whose radius is 7 meters and whose height is 15 meters.

13. _____

Perform the following unit conversions.

14. 83 in.^{2} = _____ ft^2

14. _____

15. 109 cm^2 = _____ in.^{2}

15. _____

16. 356 m^3 = _____ yd^3

16. _____

17. 900 ft^3 = _____ mi^3

17. _____

CONGRUENT
AND
SIMILAR TRIANGLES;
CONSTRUCTIONS;
AND
COORDINATE GEOMETRY

6.1 CONGRUENT TRIANGLES

OBJECTIVE

> **1** State the minimum requirements for triangles to be congruent; write correspondences for congruent triangles.

CONGRUENT TRIANGLES

If the vertices of two triangles can be put into a correspondence (can be paired) in such a way that the corresponding angles and sides of the triangle are congruent, then the triangles are congruent and the pairing is called a **congruency correspondence**.

In Figure 6.1 the triangles are congruent because the correspondence that pairs vertices A and P, B and Q, and C and R yields a congruency correspondence between the angles and sides of the triangle. Therefore,

$$\triangle ABC \cong \triangle PQR$$

FIGURE 6.1

Triangle ABC is congruent to triangle PQR

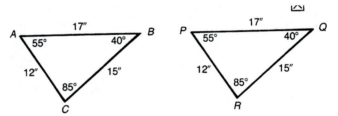

In giving the congruency statement between triangles ABC and PQR, the first triangle could have been named in any order of its vertices. But once the first triangle has been named, the second triangle must be named with its vertices in the correspondence order.

If $\triangle RST \cong \triangle VWZ$, then the correspondence that pairs R to V, S to W, and T to Z is the congruency correspondence. Notice that you can determine the congruency correspondence between the angles and sides of the triangles without drawing the triangle. If $\triangle RST \cong \triangle VWZ$, then $\angle R \cong \angle V$, $\angle S \cong \angle W$, $\angle T \cong \angle Z$, $\overline{RS} \cong \overline{VW}$, $\overline{ST} \cong \overline{WZ}$, and $\overline{RT} \cong \overline{VZ}$.

The corresponding parts (sides and angles) of two triangles may be marked congruent without actually showing the measures of the corresponding sides and angles. In Figure 6.2 the markings indicate that $\angle X \cong \angle U$, $\angle Y \cong \angle V$, $\angle Z \cong \angle W$, $\overline{XY} \cong \overline{UV}$, $\overline{XZ} \cong \overline{UW}$, and $\overline{YZ} \cong \overline{VW}$.

FIGURE 6.2

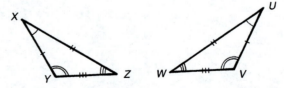

CORRESPONDING PARTS OF CONGRUENT TRIANGLES

Two triangles are congruent iff their corresponding parts are congruent.

If you know that all of the corresponding parts of one triangle are congruent to all of the corresponding parts of another triangle, then you know that the triangles are congruent. But is it possible to know that just some of their parts are congruent and still be able to conclude that the triangles are congruent? The triangles in Figure 6.3a have two pairs of corresponding angles congruent, but the triangles are clearly not congruent. The triangles in Figure 6.3b have two pairs of corresponding sides congruent and one pair of angles congruent, but the triangles are not congruent.

FIGURE 6.3

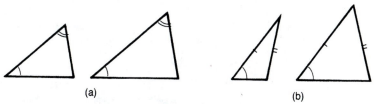

(a) (b)

At this point you may think that for triangles to be congruent all of their corresponding parts have to be congruent. There are several situations in which you can determine that two triangles are congruent without first knowing that all of their corresponding parts are congruent. Three of these situations are summarized as follows.

RULES FOR CONGRUENT TRIANGLES

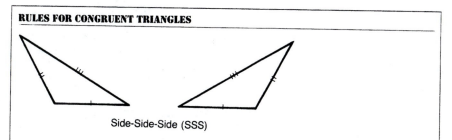

Side-Side-Side (SSS)

If three sides of one triangle are congruent to three sides of another triangle, the triangles are congruent.

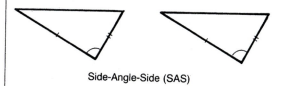

Side-Angle-Side (SAS)

If two sides and the **included** angle (angle located between the two sides) of one triangle are congruent to two sides and the included angle of another triangle, the triangles are congruent.

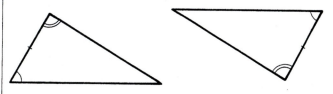

Angle-Side-Angle (ASA)

If two angles and the included side of one triangle are congruent to two angles and the included side of another triangle, the triangles are congruent.

The Rules for Congruent Triangles can be used to show that two triangles are congruent; therefore, by the definition of congruent triangles it follows that all of the corresponding parts of the two triangles are congruent.

EXAMPLE 1   The two triangles shown have congruent parts as marked. Can you conclude that angles A and D are congruent?

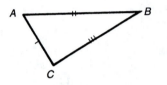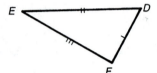

Solution   Triangles ABC and DEF are congruent by the SSS Rule for Congruent Triangles. Since corresponding parts of congruent triangles are congruent, it follows that $\angle A \cong \angle D$.

EXAMPLE 2   For Problems 1–3, the corresponding parts of the two triangles are congruent as marked. Using the Rules for Congruent Triangles, show how the two triangles are congruent.

1. Consider $\triangle ABC$ and $\triangle CDA$.   2. Consider $\triangle QOP$ and $\triangle ROS$.

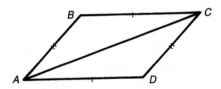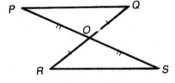

3. Consider $\triangle ABC$ and $\triangle DCB$.

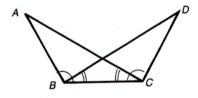

Solution   1. $\triangle ABC$ and $\triangle CDA$ share a common side $\overline{AC}$. Thus we know that

$$\overline{AB} \cong \overline{CD}$$
$$\overline{BC} \cong \overline{DA}$$
$$\overline{AC} \cong \overline{CA}$$

Therefore the triangles are congruent by the SSS Rule.

2. $\angle QOP \cong \angle SOR$ because they are vertical angles. Therefore,

$$\overline{PO} \cong \overline{SO}$$
$$\angle POQ \cong \angle SOR$$
$$\overline{RO} \cong \overline{QO}$$

Thus the triangles are congruent by the SAS Rule.

3. $\triangle ABC$ and $\triangle DCB$ share common side $\overline{BC}$. Therefore,

$$\angle DBC \cong \angle ACB$$
$$\overline{BC} \cong \overline{CB}$$
$$\angle ABC \cong \angle DCB$$

Thus the triangles are congruent by the ASA Rule.   ◀

QUICK CHECK

Answer each of the following true or false.

1. If $\triangle ABC \cong \triangle PQR$, then $\overline{AC} \cong \overline{PR}$.

2. Triangles can be shown congruent by the SSA Rule.

3. Congruent means "exactly the same set of points."

4. Triangles can be shown congruent by the AAA Rule.

5. Corresponding parts of congruent triangles are congruent.

ANSWERS

1. True 2. False 3. False 4. False 5. True

A

In Exercises 1–6 the triangles are congruent as marked. Name a congruency correspondence between each pair of triangles.

1.
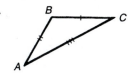 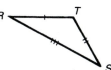

1. _____

2.
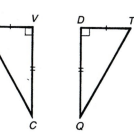

2. _____

3.
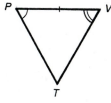

3. _____

4.
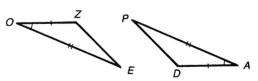

4. _____

5.
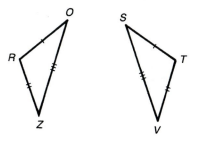

5. _____

6.
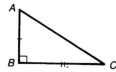

6. _____

For each of the following congruency statements, list the corresponding congruent parts.

7. $\triangle RST \cong \triangle UVW$.

8. $\triangle PQR \cong \triangle DEF$.

9. $\triangle PQR \cong \triangle SRQ$.

10. $\triangle UVW \cong \triangle RST$.

11. $\triangle ABC \cong \triangle RST$.

12. $\triangle ABC \cong \triangle CAB$.

7. _____

8. _____

9. _____

10. _____

11. _____

12. _____

The pairs of triangles in Exercises 13–18 are congruent by one of the Rules for Congruent Triangles. Name the rule and list the congruent parts.

13.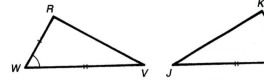

13. _____

14.

14. _____

15.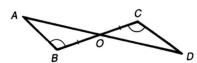

15. _____

16.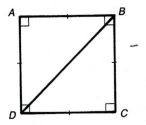

16. _____

17.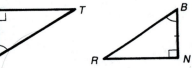

17. _____

18.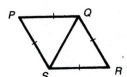

18. _____

B

19. If $\triangle ABC \cong \triangle CBA$, what kind of triangle is $\triangle ABC$?

20. If $\triangle ABC \cong \triangle CAB$, what kind of triangle is $\triangle ABC$?

21. A diagonal divides a parallelogram into two congruent triangles. Why?

22. The median to the base divides an isosceles triangle into two congruent triangles. Why? (The base of an isosceles triangle is the noncongruent side.)

23. The diagonals of a rhombus divide it into four congruent triangles. Why?

24. If two angles of one triangle are congruent to two angles of another triangle, then their third pair of corresponding angles must be congruent. Why?

25. If point P is any point on the perpendicular bisector of $\overline{AB}$, then $AP = BP$. Why?

19. _____

20. _____

21. _____

22. _____

23. _____

24. _____

25. _____

6.2 SIMILAR TRIANGLES

OBJECTIVE

▶1 State the requirements for similar triangles; write proportionality statements for corresponding parts of similar triangles; use proportionality between similar triangles to find side lengths.

1▶ SIMILAR TRIANGLES

Two or more pairs of numbers are **proportional** if and only if there exists a single real number k, called the **proportionality constant**, such that the second number in each pair can be obtained by multiplying k and the first number of the pair. For example, we can say that 4, 7, and 9 are proportional to 12, 21, and 27, respectively, because

$$12 = 3 \cdot 4$$
$$21 = 3 \cdot 7$$
$$27 = 3 \cdot 9$$

In this case the proportionality constant is 3.

DEFINITION OF PROPORTIONAL

If A is proportional to B, then there exists a real number k such that

$$A = kB$$

k is called the proportionality constant.

Two triangles are **similar** if and only if their corresponding angles are congruent and the measures of their corresponding sides are proportional. That is, similar triangles have the same shape but may have different sizes. The two pairs of triangles in Figure 6.4 are similar.

FIGURE 6.4
Two pairs of similar triangles

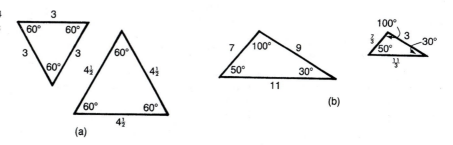

The symbol $\sim$ means "is similar to." Thus, to indicate that triangle ABC and PQR are similar, we write $\triangle ABC \sim \triangle PQR$, which is read "triangle ABC is similar to triangle PQR."

From the definition of similar triangles, if $\triangle ABC \sim \triangle PQR$, then the measures of their corresponding sides are proportional. Therefore, a single real number k exists such that

$$AB = k(PQ)$$
$$BC = k(QR)$$
$$AC = k(PR)$$

Suppose $\triangle ABC \sim \triangle PQR$, and you are asked to find the unknown lengths in the figure shown in the margin.

Since the triangles are similar, we know that there exists a single real number k such that

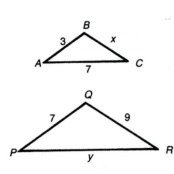

(1) $3 = k \cdot 7$

(2) $7 = k \cdot y$

(3) $x = k \cdot 9$

Equation (1) can be used to find the value of k.

$$3 = 7k$$
$$\frac{3}{7} = k$$

Now we can substitute the proportionality constant into equations (2) and (3) to determine the values of x and y. Substituting $\frac{3}{7}$ for k in equation (2) yields

$$7 = \left(\frac{3}{7}\right)y$$
$$\left(\frac{7}{3}\right)7 = y$$
$$\frac{49}{3} = y$$

Substituting $\frac{3}{7}$ for k in equation (3)

$$x = k \cdot 9$$
$$x = \left(\frac{3}{7}\right)(9)$$
$$x = \frac{27}{7}$$

The following rule states that two triangles are similar under certain conditions.

A A RULE FOR SIMILAR TRIANGLES

Two triangles are similar if and only if two angles of one of the triangles are congruent to two angles of the other triangle.

E X A M P L E 3   Refer to the figure and show that $\triangle ABC \sim \triangle PBQ$.

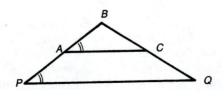

Solution   The figure indicates that $\angle BAC$ is congruent to $\angle BPQ$. Since $\triangle ABC$ and $\triangle PBQ$ share $\angle B$, they have a second pair of congruent angles. Thus

$$\angle ABC \cong \angle PBQ$$

and

$$\angle BAC \cong \angle BPQ$$

Therefore, the triangles are similar by the AA Rule for Similar Triangles.   ◀

EXAMPLE 4   In the figure shown, if $\triangle ABC \sim \triangle PBQ$, find the unknown lengths.

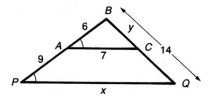

Solution   Since the triangles are similar, their corresponding sides are proportional. Thus

(1) $AB = k(PB)$ so $6 = k(15)$

(2) $AC = k(PQ)$ so $7 = k(x)$

(3) $BC = k(BQ)$ so $y = k(14)$

Solving equation (1) for k yields

$$\frac{6}{15} = k$$

$$\frac{2}{5} = k$$

Substituting $\frac{2}{5}$ for k in equation (2) yields

$$7 = \left(\frac{2}{5}\right)x$$

$$\left(\frac{5}{2}\right)7 = x$$

$$\frac{35}{2} = x$$

Substituting $\frac{2}{5}$ for k in equation (3) results in

$$y = \left(\frac{2}{5}\right)(14)$$

$$y = \frac{28}{5}$$

   ◀

EXAMPLE 5   For Problems 1 and 2, show that the given triangles are similar and then find the unknown lengths.

1.

2.

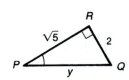

Solution   1. $\angle DOC \cong \angle BOA$ because they are vertical angles. The drawing indicates that $\angle OAB \cong \angle OCD$. Thus the triangles are similar by the AA Rule for Similar Triangles.

Since the triangles are similar, their corresponding sides are proportional.

(1) $DO = k(BO)$ so $y = k(2)$

(2) $OC = k(OA)$ so $10 = k(4)$

(3) $DC = k(BA)$ so $5 = k(x)$

Solving equation (2) for k results in

$$\frac{10}{4} = k$$

$$\frac{5}{2} = k$$

Substituting $\frac{5}{2}$ for k in equation (1) yields

$$y = \left(\frac{5}{2}\right)2$$

$$y = 5$$

and substituting $\frac{5}{2}$ for k in equation (3) yields

$$5 = \left(\frac{5}{2}\right)x$$

$$\left(\frac{2}{5}\right)5 = x$$

$$2 = x$$

2. Angles B and R are congruent because they are both marked as right angles. Angles C and P are marked congruent. Therefore, the triangles are similar by the AA Rule for Similar Triangles.

Since $\triangle ABC \sim \triangle QRP$, their corresponding sides are proportional. Therefore

(1) $AB = k(QR)$ so $6 = k(2)$

(2) $BC = k(RP)$ so $x = k(\sqrt{5})$

(3) $AC = k(QP)$ so $9 = k(y)$

Solving equation (1) for k gives

$$\frac{6}{2} = k$$

$$3 = k$$

Substituting 3 for k in equation (2) yields

$$x = 3(\sqrt{5})$$

$$x = 3\sqrt{5}$$

Substituting 3 for k in equation (3) yields

$9 = 3y$

$3 = y$

◀

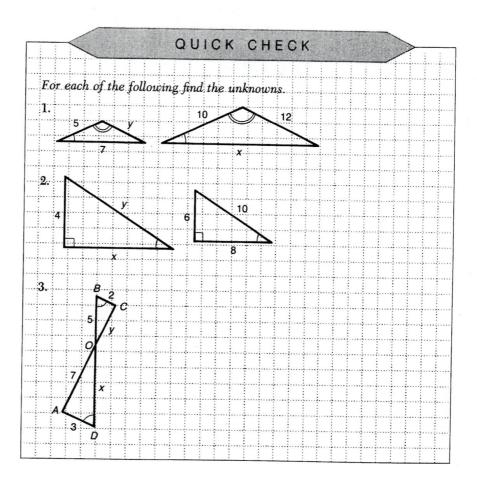

QUICK CHECK

For each of the following find the unknowns.

1.

2.

3.

A

Write the similarity correspondence for each pair of similar triangles.

1.

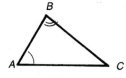

1. _____

2.

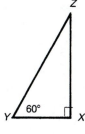

2. _____

3.

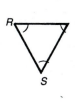

3. _____

4.

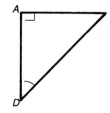

4. _____

5.

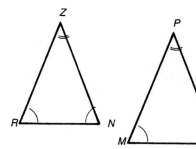

5. _____

6.

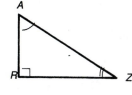

6. _____

7. If $\triangle ABC \sim \triangle PQR$, $AB = 7$, $PQ = 14$, and $AC = 8$, find PR.

8. If $\triangle RST \sim \triangle UVW$, $RT = 5$, $UW = 4$, $ST = 7$, and m $\cdot \angle RTS = 40°$, find VW and m $\angle UWV$.

9. If $\triangle ABP \sim \triangle XYZ$, $AB = 4$, $XY = 9$, and $YZ = 5$, find BP.

10. If $\triangle AOR \sim \triangle BOT$, $AR = 40$, $BT = 60$, and $OR = 20$, find OT.

11. If $\triangle ZTN \sim \triangle WXY$, $TN = 5.8$, $XY = 9.1$, and $ZN = 4.1$, find WY.

12. In the figure shown $\overline{AB} \parallel \overline{DE}$. Find two pairs of congruent angles in triangles ABC and DEC.

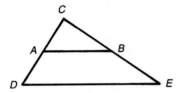

13. In the figure shown $\overline{AB} \parallel \overline{CD}$. Find two pairs of congruent angles in triangles AOB and DOC.

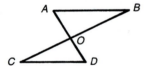

14. If two right triangles have a single pair of acute angles congruent, then they are similar. Why?

15. In the figure shown, name three triangles that are similar to each other.

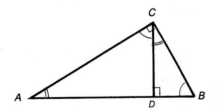

16. Using the results in Exercise 15, find each unknown.

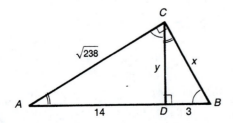

For each of the following, find the unknowns.

17.

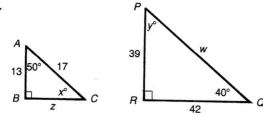

17. _____

18.

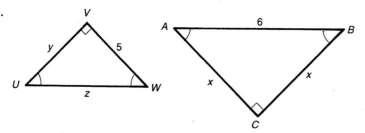

18. _____

19.

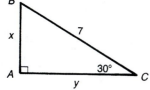

19. _____

20.

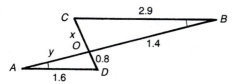

20. _____

21.

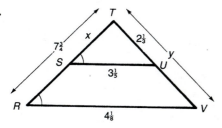

21. _____

22.

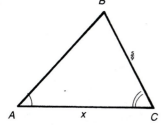

22. _____

23.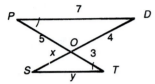

23. _____

24.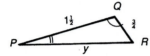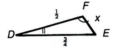

24. _____

25.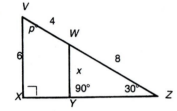

25. _____

B

Answer each of the following true or false.

26.   All equilateral triangles are similar.

27.   All 30–60–90 triangles are similar.

28.   All 45–45–90 triangles are similar.

29.   All isosceles triangles are similar.

30.   All scalene triangles are similar.

26. _____

27. _____

28. _____

29. _____

30. _____

6.3 CONSTRUCTIONS

OBJECTIVE

1 ▶ Draw the 13 fundamental constructions of geometry and use these constructions to draw more complex constructions.

1 ▶ FUNDAMENTAL CONSTRUCTIONS OF GEOMETRY

In mathematics a **construction** is a drawing that is made using only a pencil, paper, a compass, and an unmarked straightedge. By "unmarked" we mean that you cannot measure feet, inches, and so on.

In the field of mechanical drawing, draftsmen use a compass and straightedge daily in their work. Our everyday lives are filled with situations that require scaled drawings called schematics. Schematics range from blueprints of houses and commercial buildings to elaborate drawings of a bridge to a simple drawing for making a hexagonal bolt.

In this section we study 13 fundamental constructions and then see how these constructions can be adapted and modified to make more complex constructions.

The figure in the margin shows the parts of a compass. To use a compass efficiently, there are several guidelines to keep in mind.

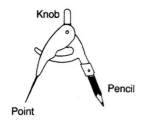

Knob

Pencil

Point

1. Keep your pencil sharp.
2. Always hold the compass by the knob when marking arcs.
3. Put pressure on the point rather than the pencil when marking arcs.
4. Slightly tilt the compass and drag the pencil in the direction of the tilt.

Marking an arc

The following terminology is commonly used when describing how to make a construction.

1. "Mark an arc" means to draw an arc using your compass.
2. "Measure with your compass" means to set your compass to a setting specified by two given points. For example, to measure a line segment, put the point of the compass on one endpoint of the line segment and set the compass so that the pencil point exactly touches the other endpoint.
3. "Prime letters" are letters with marks such as A', B', and C', which are read "A prime," "B prime," and "C prime." Prime letters are often used to name points in a construction that correspond to given letters in the figure. For example, if you are to construct a line segment congruent to a given line segment AB, your constructed line segment is designated line segment $A'B'$.

Now let's draw the 13 fundamental constructions of geometry and see how these constructions can be modified to make more complex constructions.

Construction of a Line Segment Congruent to a Given Line Segment

Given line segment:

1. Draw a ray using your straightedge.

2. Measure the given line segment with your compass.

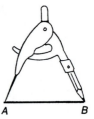

3. Being careful not to let your compass slip, put the point on the endpoint of the ray and mark an arc intersecting the ray.

4. Label the endpoint of the ray A' and the point where the arc intersects the ray B'.

Thus $\overline{A'B'} \cong \overline{AB}$.

Construction of a Midpoint and a Perpendicular Bisector of a Given Line Segment

Given line segment:

1. Set your compass to a setting that is more than one-half the length of the given line segment.

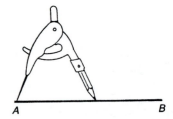

2. Put the point of the compass on one endpoint and mark arcs above and below the line segment.

3. Put your compass point on the other endpoint and mark arcs above and below the line segment. These arcs intersect the arcs drawn in Step 2.

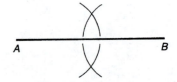

4. Place your straightedge along the line determined by the points of intersection of the arcs and draw line *l*. Label the point where line *l* intersects the line segment *M*.

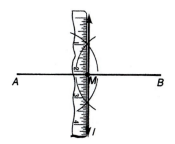

Thus point *M* bisects $\overline{AB}$ and line *l* is the perpendicular bisector of $\overline{AB}$.

Construction of an Angle Congruent to a Given Angle

Given angle:

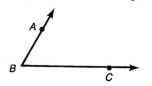

1. Using your straightedge, draw a ray located at the place you want one side of your angle to be.

2. Set your compass to any setting you choose, put the compass point on the vertex of the given angle, and mark an arc cutting both sides of the given angle.

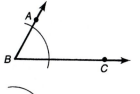

3. Put your compass point on the endpoint of the ray constructed in Step 1. Mark an arc intersecting the ray that is at least as long as the arc drawn in Step 2. This arc will intersect the interior of your constructed angle. Label the point where the arc intersects the ray *C'* and the endpoint of the ray *B'*.

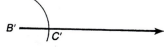

4. Using your compass, measure the width of the opening of the given angle by measuring the distance between the points of intersection of the arc and the given angle in Step 2.

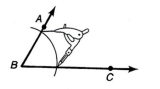

5. Put your compass point on *C'* and mark an arc intersecting the arc you constructed in Step 3. Label this point of intersection *A'*.

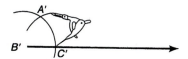

6. With your straightedge, construct $\overrightarrow{B'A'}$.

Thus $\angle A'B'C' \cong \angle ABC$.

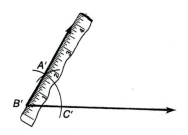

Construction of the Bisector of a Given Angle

Given angle:

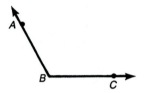

1. Place the point of your compass on the vertex of the given angle and mark an arc intersecting both sides of the angle. Label the points where the arc intersects the angle D and E.

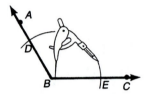

2. Put the compass point on D and mark an arc in the interior of the angle as shown.

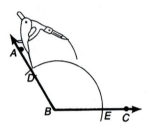

3. Being careful not to let your compass slip, place your compass point on E and mark an arc intersecting the arc drawn in Step 2. Label the point where the arcs intersect F.

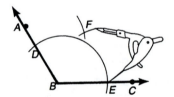

4. Using your straightedge, draw $\overrightarrow{BF}$.

Thus $\overrightarrow{BF}$ bisects $\angle ABC$

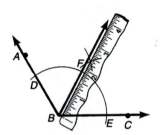

EXAMPLE 6   Construct an angle whose measure is 45°.

Solution   The strategy is to construct a right angle and then bisect it to get two 45° angles. To construct a right angle, we construct the perpendicular bisector of a line segment:

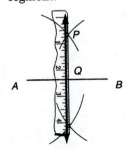

Since $\overrightarrow{PQ} \perp \overline{AB}$, $\angle PQB$ is a right angle. Now let's bisect $\angle PQB$:

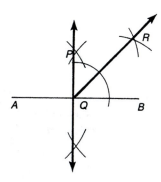

Therefore, m $\angle RQB = 45°$. ◀

Construction of a Triangle That Is Congruent to a Given Triangle by the SSS Rule for Congruent Triangles

Given triangle:

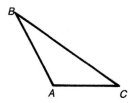

1. Construct a line segment congruent to any chosen side of the given triangle. Let's use $\overline{AC}$.
2. Measure $\overline{AB}$ with your compass, put the compass point on A', and mark an arc on one side of $\overrightarrow{A'C'}$.

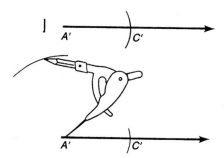

3. Measure $\overline{BC}$ with your compass, put your point on C', and mark an arc intersecting the arc drawn in Step 2. Label the point where the arcs intersect B'.

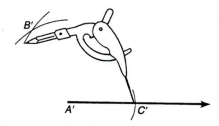

4. Using your straightedge, draw $\overline{A'B'}$ and $\overline{B'C'}$.

Therefore, $\triangle A'B'C' \cong \triangle ABC$ by SSS.

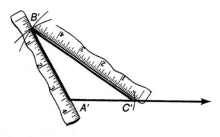

E X A M P L E 7   Construct an equilateral triangle whose sides are the length of the given line
segment.

Given line segment: A ———— B

Solution   1. Construct a line segment congruent
to $\overline{AB}$.

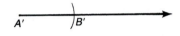

2. Without changing your compass setting, put the compass point on A'
and mark an arc on one side of $\overrightarrow{A'B'}$.

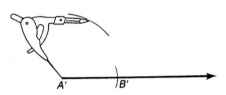

3. Without changing your compass setting, put the compass point on B'
and mark an arc intersecting the arc
drawn in Step 2. Label the point
where the arcs intersect C'.

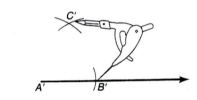

4. Using your straightedge, draw $\overline{A'C'}$
and $\overline{B'C'}$.

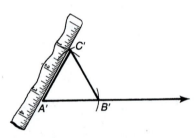

Thus $\triangle A'B'C'$ is equilateral and each side is congruent to $\overline{AB}$. Note that we not
only constructed an equilateral triangle but we also constructed three 60° angles. ◀

Construction of a Triangle That Is Congruent to a Given Triangle by the SAS Rule for
Congruent Triangles

Given triangle:

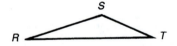

1. Construct an angle congruent to any
angle of the triangle. Let's use ∠R and
name it ∠AR'B.

2. Measure $\overline{RS}$ with your compass, put
the compass point on R', and mark an
arc on $\overrightarrow{R'A}$. Name the point where the
arc intersects $\overrightarrow{R'A}$ point S'.

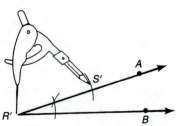

3. Measure $\overline{RT}$ with your compass, put the compass point on R', and mark an arc on $\overrightarrow{R'B}$. Name the point where the arc intersects $\overrightarrow{R'B}$ point T'.

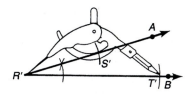

4. Using your straightedge, draw $\overline{S'T'}$.

Thus $\triangle R'S'T' \cong \triangle RST$ by SAS.

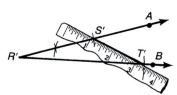

Construction of a Triangle That Is Congruent to a Given Triangle by the ASA Rule for Congruent Triangles

Given triangle:

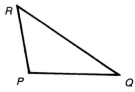

1. Construct a line segment congruent to one side of the given triangle. Let's use $\overline{PQ}$.

2. Construct an angle with vertex at P' and congruent to $\angle P$. Use $\overrightarrow{P'Q'}$ as one side of the angle.

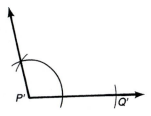

3. Construct an angle with vertex at Q' and congruent to $\angle Q$. Use $\overrightarrow{Q'P'}$ as one side of the angle. Label the point where the rays intersect R'.

Thus $\triangle P'Q'R' \cong \triangle PQR$ by ASA.

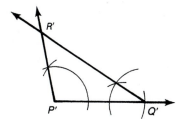

Construction of a Line Parallel to a Given Line Containing a Given Point

Given line and point: $P \bullet$

1. Draw a line through point P intersecting line l. Label the point of intersection Q. Label points R and S as shown.

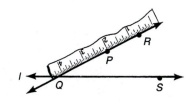

2. Construct an angle congruent to ∠PQS having vertex P and $\overrightarrow{PR}$ as one of its sides. Label point T as shown.

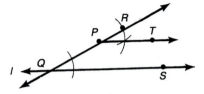

3. Draw $\overleftrightarrow{PT}$.

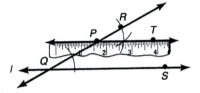

Thus $\overleftrightarrow{PT} \parallel$ line l because corresponding angles RPT and PQS are congruent.

Construction of a Line Perpendicular to a Given Line through a Point on the Given Line

Given line and point:

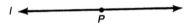

1. Put the compass point on point P and mark an arc on each side of P. Be careful not to let your compass slip. Label points Q and R as shown.

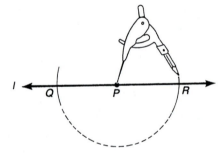

2. Lengthen your compass slightly and put the compass point on Q and mark an arc on either side of line l.

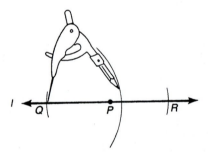

3. Being very careful not to let your compass slip, put the compass point on R and mark arcs intersecting the arcs drawn in Step 2. Label the points of intersection A and B as shown.

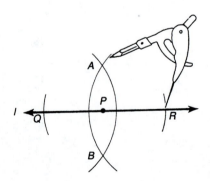

4. Using your straightedge, draw $\overleftrightarrow{AB}$. Point P should be on $\overleftrightarrow{AB}$. If it is not, you have probably let your compass slip.

Thus $\overleftrightarrow{AB}$ contains P and $\overleftrightarrow{AB} \perp$ line l.

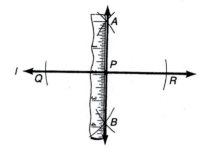

Construction of a Line Perpendicular to a Given Line and Containing a Point Not on the Given Line

Given line and point:

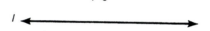

1. Set your compass to a setting so that when you put the point on P it will mark an arc that intersects line l twice. Label the points of intersection R and S.

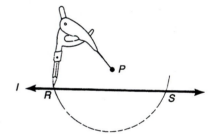

2. Set your compass to any setting that is more than one-half $\overline{RS}$. Place the point on R and mark an arc on the side of the line opposite P.

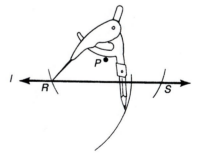

3. Being careful not to let your compass slip put the compass point on S and mark an arc intersecting the arc drawn in Step 3. Label this point of intersection T.

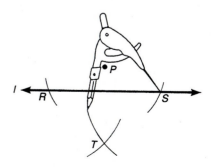

4. With your straightedge draw $\overleftrightarrow{PT}$.
Thus $\overleftrightarrow{PT}$ contains P and $\overleftrightarrow{PT} \perp$ line l.

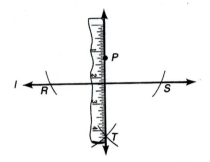

Construction of a Set of Points That Divides a Given Line Segment into Three Congruent Parts

Given line segment:

1. Choosing either endpoint of the line segment and using your straightedge, draw a ray sharing an endpoint with the line segment. Let's draw $\overrightarrow{AC}$.

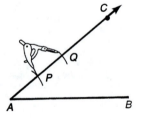

2. With your compass set to any length, put your point on A and mark an arc intersecting $\overrightarrow{AC}$. Label the point of intersection P.

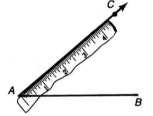

3. Being careful not to let your compass slip, put the compass point on P and mark another arc intersecting $\overrightarrow{AC}$. Label the point of intersection Q.

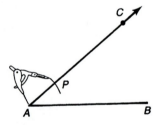

4. Put the compass point on Q and mark a third arc intersecting $\overrightarrow{AC}$. Label the point of intersection R.

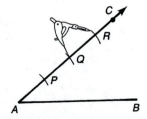

5. Using your straightedge, draw $\overline{RB}$.

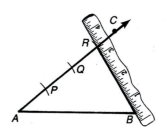

6. Construct an angle congruent to $\angle ARB$ having vertex at Q and one side $\overrightarrow{QA}$. Label the point of intersection of the constructed angle's side and $\overline{AB}$ point S.

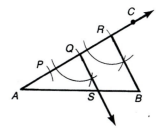

7. Construct another angle congruent to $\angle ARB$ having vertex at P and one side $\overrightarrow{PA}$. Label the point of intersection of the constructed angle's side and $\overline{AB}$ point T as shown.

Thus $\overline{AT} \cong \overline{TS} \cong \overline{SB}$.

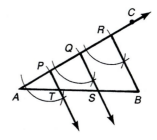

This construction can be modified slightly to divide a line segment into any number of congruent parts. For example, we can use the following steps to divide the line segment into four congruent parts.

1. Mark off four arcs on $\overrightarrow{AC}$. Label points of intersection P, Q, U, and W, respectively.
2. Connect the last arc's intersection to B.
3. Construct three angles congruent to $\angle AWB$ having vertices U, Q, and P intersecting $\overline{AB}$ at points T, S, and R.

Thus $\overline{AR} \cong \overline{RS} \cong \overline{ST} \cong \overline{TB}$.

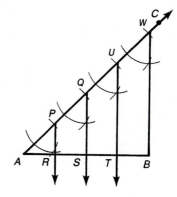

Construction of a Circle through Three Given Noncollinear Points

Given noncollinear points: • A   • B

                                                   • C

1. Using your straightedge, draw $\overline{AB}$ and then construct its perpendicular bisector.

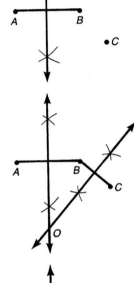

2. Using your straightedge, draw $\overline{BC}$ and then construct its perpendicular bisector. Label the point where the perpendicular bisectors intersect O.

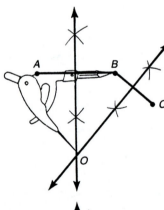

3. Measure the distance from O to one of the given points A, B, or C. (The distances from O to A, O to B, and O to C should all be the same.)

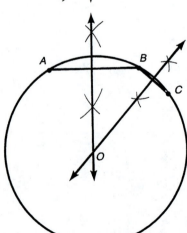

4. Being careful not to let your compass slip, put the compass point on O and draw a circle.

Thus circle O contains noncollinear points A, B, and C.

Construction of a Regular Hexagon

1. On your paper mark a point and label it O.

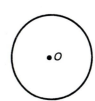

2. Using your compass, draw circle O. Don't change your compass setting because we are going to use it again in Steps 4 through 7.

3. Label any point on the circle point A.

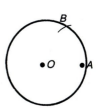

4. Put the compass point on A, mark an arc intersecting the circle, and label the point of intersection B.

5. Put the compass point on B, mark another arc intersecting the circle, and label the point of intersection C.

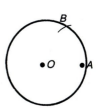

6. Put the compass point on C, mark another arc intersecting the circle, and label the point of intersection D.

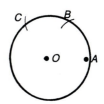

7. Continue the process described in Steps 4 through 6 until you have marked off six arcs. The last arc should intersect point A.

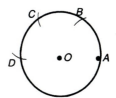

8. Using your straightedge, draw $\overline{AB}$, $\overline{BC}$, $\overline{CD}$, $\overline{DE}$, $\overline{EF}$, and $\overline{FA}$.

Thus polygon $ABCDEF$ is a regular hexagon.

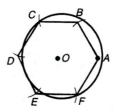

A

Draw the specified constructions.

ANSWERS

1. A line segment congruent to $\overline{PQ}$.

1. _____

P ————————————— Q

2. An angle congruent to $\angle ABC$.

2. _____

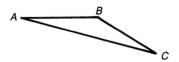

3. The bisector of $\angle PQR$.

3. _____

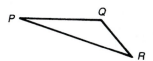

4. The perpendicular bisector of $\overline{RS}$.

4. _____

R ——————— S

5. A triangle congruent to $\triangle ABC$ by SSS.

5. _____

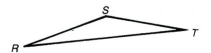

6. A triangle congruent to $\triangle PQR$ by ASA.

6. _____

7. A triangle congruent to $\triangle RST$ by SAS.

7. _____

8.   A line parallel to line l containing point P.

8. _____

9.   A line perpendicular to line l containing point P.

9. _____

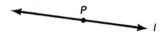

10.   A line perpendicular to line l containing point P.

10. _____

11.   Divide $\overline{AB}$ into five congruent segments.

11. _____

12.   A circle containing point P, Q, and R.

12. _____

$P \bullet$ $\bullet Q$

$\bullet R$

13.   A regular hexagon.

13. _____

B

14.   A 30° angle.

14. _____

15.   A 15° angle.

15. _____

16.   The medians in $\triangle ABC$.

16. _____

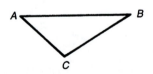

17.   The altitudes in $\triangle ABC$.

17. _____

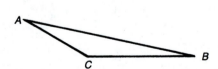

18.   A 30–60–90 triangle.

18. _____

19.   An isosceles right triangle.

19. _____

20.   A regular octagon.

20. _____

6.4 COORDINATE GEOMETRY

OBJECTIVES

▶ **1** Find the distance between two points with known coordinates.

▶ **2** Find the coordinates of the midpoint of a given line segment.

▶ **3** Use coordinates to verify geometrical statements concerning sets of points in the Cartesian plane.

▶ **1** THE DISTANCE FORMULA

The distance between any two points in the Cartesian plane can be calculated if the points have known coordinates. Let's consider points A and B having coordinates (x_1, y_1) and (x_2, y_2), respectively. In Figure 6.5 we have drawn $\overline{AB}$ and constructed a right triangle whose hypotenuse is $\overline{AB}$ and whose legs are parallel to the x-axis and to the y-axis, respectively. We see from Figure 6.5 that point C must be located at (x_2, y_1) and that the lengths of line segments $\overline{AC}$ and $\overline{BC}$ are $x_2 - x_1$ and $y_2 - y_1$, respectively. Now using the Pythagorean Theorem,

$$(AB)^2 = (AC)^2 + (BC)^2$$
$$(AB)^2 = (x_2 - x_1)^2 + (y_2 - y_1)^2$$
$$\sqrt{(AB)^2} = \sqrt{(x_2 - x_1)^2 + (y_2 - y_1)^2}$$

Thus $AB = \sqrt{(x_2 - x_1)^2 + (y_2 - y_1)^2}$.

FIGURE 6.5

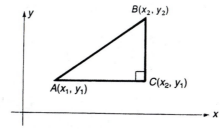

The formula we have just obtained for finding the distance between two points A and B is called the **Distance formula**. The Distance formula can be used to find the distance between any two points having known coordinates.

THE DISTANCE FORMULA

The distance d between two points whose coordinates are (x_1, y_1) and (x_2, y_2), respectively, is given by the formula

$$d = \sqrt{(x_2 - x_1)^2 + (y_2 - y_1)^2}$$

EXAMPLE 8   Find the distance between two points in the Cartesian plane whose coordinates are $(-3, 5)$ and $(2, -7)$.

Solution   Let $(x_1, y_1) = (-3, 5)$ and $(x_2, y_2) = (2, -7)$. Then the distance d between the points is

$$d = \sqrt{(x_2 - x_1)^2 + (y_2 - y_1)^2}$$
$$= \sqrt{[2 - (-3)]^2 + (-7 - 5)^2}$$
$$= \sqrt{(5)^2 + (-12)^2}$$
$$= \sqrt{25 + 144}$$
$$= \sqrt{169}$$
$$= 13 \text{ units}$$

QUICK CHECK

For each of the following find the distance between each pair of points.

1. $(-5, 6), (-7, -3)$ **2.** $(4, 8), (5, 8)$ **3.** $(-6, 7), (-6, 2)$

---

2 ▶ THE MIDPOINT FORMULA

The coordinates of the midpoint (bisector) of a line segment can be calculated once the coordinates of the line segment's endpoints are known. In Figure 6.6 $\overline{AB}$ has endpoints at (x_1, y_1) and (x_2, y_2). Since the coordinates of midpoint M are unknown, they are represented by (x, y). Right triangles ABC, AMD, and MBE are drawn and the coordinates for D and E are obtained by examining the resulting figure. Three pairs of similar triangles are formed as follows.

1. $\triangle ABC \sim \triangle AMD$ by AA. They share $\angle A$ and both have right angles.
2. $\triangle ABC \sim \triangle MBE$ by AA. They share $\angle A$ and both have right angles.
3. $\triangle AMD \sim \triangle MBE$ by AA. They have congruent angles because they are both similar to $\triangle ABC$.

FIGURE 6.6

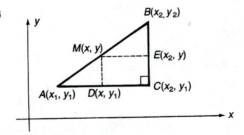

Using these similar triangle relationships, let's see if we can find an expression for x and y using the known coordinates of the endpoints.

1. $AM = k(AB)$ Since $\triangle AMD \sim \triangle ABC$

2. $\dfrac{AM}{AB} = k$ Solving the equation in Step 1 for k

3. $AM = \frac{1}{2}AB$ Since M is the midpoint of $\overline{AB}$

4. $\dfrac{AM}{AB} = \dfrac{1}{2}$ Multiplying both sides of the equation in Step 3 by $\dfrac{1}{AB}$

5. $k = \dfrac{1}{2}$ Substituting $\dfrac{1}{2}$ for $\dfrac{AM}{AB}$ in Step 2 from Step 4

6. $AD = k(AC)$ Since $\triangle AMD \sim \triangle ABC$

7. $AD = \frac{1}{2}AC$ Substituting $\frac{1}{2}$ for k in Step 6 from Step 5

Now we substitute $x - x_1$ for AD and $x_2 - x_1$ for AC in Step 7 and solve the resulting equation for x.

8. $x - x_1 = \frac{1}{2}(x_2 - x_1)$

9. $x - x_1 = \frac{1}{2}x_2 - \frac{1}{2}x_1$ Distributive Property

10. $x = \frac{1}{2}x_2 - \frac{1}{2}x_1 + x_1$ Adding x_1 to both sides

11. $x = \frac{1}{2}x_2 + \frac{1}{2}x_1$ Adding like terms

12. $x = \dfrac{x_2 + x_1}{2}$ Distributive Property and Definition of Division

Similarly, using the fact that $\triangle ABC$ and $\triangle MBE$ are similar, it can be shown that

$$y = \frac{y_2 + y_1}{2}$$

Thus the coordinates of the midpoints of a line segment are as follows.

THE MIDPOINT FORMULA

The coordinates of the midpoint of a line segment whose endpoints are (x_1, y_1) and (x_2, y_2) are

$$\left(\frac{x_1 + x_2}{2}, \frac{y_1 + y_2}{2} \right)$$

EXAMPLE 9   Find the midpoint for a line segment whose endpoints are located at $(-7, -5)$ and $(2, -5)$.

Solution   Using the Midpoint formula, let $(x_1, y_1) = (-7, -5)$ and $(x_2, y_2) = (2, -5)$. Then the abscissa x for the midpoint is

$$x = \frac{x_1 + x_2}{2}$$

$$= \frac{-7 + 2}{2}$$

$$= \frac{-5}{2}$$

$$= -2\frac{1}{2}$$

and the ordinate y for the midpoint is

$$y = \frac{y_1 + y_2}{2}$$

$$= \frac{-5 + (-5)}{2}$$

$$= \frac{-10}{2}$$

$$= -5$$

Thus the midpoint is located at $(-2\frac{1}{2}, -5)$.   ◄

EXAMPLE 10   The midpoint of a line segment is located at $(2, 3)$ and one endpoint of the line segment is at $(-7, -2)$. Find the coordinates of the other endpoint of the line segment.

Solution   Using the Midpoint formula, if we let $(x_1, y_1) = (-7, -2)$ and $(x, y) = (2, 3)$, then we need to find the coordinates of the other endpoint, that is, (x_2, y_2).

$$x = \frac{x_1 + x_2}{2}$$

$$2 = \frac{-7 + x_2}{2}$$

$$4 = -7 + x_2 \quad \text{Multiplying both sides by 2}$$

$$11 = x_2 \quad\quad\quad \text{Adding 7 to both sides}$$

and

$$y = \frac{y_1 + y_2}{2}$$

$$3 = \frac{-2 + y_2}{2}$$

$$6 = -2 + y_2$$

$$8 = y_2$$

Thus the other endpoint is located at $(11, 8)$.   ◄

QUICK CHECK

In Problems 1–3 find the coordinates of the midpoint for each line segment whose endpoints have the given coordinates.

1. $(-3, 0)$, $(4, 7)$ 2. $(0, -8)$, $(5, 0)$ 3. $(-3, 2)$, $(-3, 7)$

4. Find the coordinates of the endpoint of a line segment having a midpoint at $(-7, -3)$ and one endpoint at $(7, -11)$.

ANSWERS

1. $(\frac{1}{2}, \frac{7}{2})$ 2. $(\frac{5}{2}, -4)$ 3. $(-3, \frac{9}{2})$ 4. $(-21, 5)$

3 ▶ GEOMETRICAL FIGURES IN THE CARTESIAN PLANE

The Midpoint formula and the Distance formula can be used with geometrical figures placed in the Cartesian plane.

EXAMPLE 11

The vertices of triangle ABC are located at $(0, 0)$, $(10, 0)$, and $(5, 15)$, as shown in the figure. Show that the triangle is isosceles.

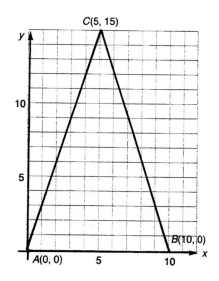

Solution

To show that the triangle is isosceles, we use the Distance formula to find the lengths of each side.

$$AB = \sqrt{(10 - 0)^2 + (0 - 0)^2}$$
$$= \sqrt{100}$$
$$= 10$$

$$AC = \sqrt{(5 - 0)^2 + (15 - 0)^2}$$
$$= \sqrt{25 + 225}$$
$$= \sqrt{250}$$
$$= 5\sqrt{10}$$

$$BC = \sqrt{(5 - 10)^2 + (15 - 0)^2}$$
$$= \sqrt{(-5)^2 + 225}$$
$$= \sqrt{25 + 225}$$
$$= \sqrt{250}$$
$$= 5\sqrt{10}$$

Since $AC = BC$, the triangle is isosceles. ◀

EXAMPLE 12   In the figure shown, the vertices of $\triangle ABC$ are located at $(2, 3)$, $(-7, 5)$ and $(4, 8)$, respectively. Find the length of the median from A to $\overline{BC}$.

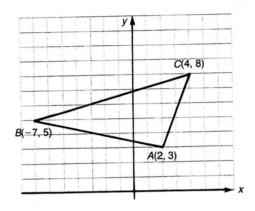

Solution   To find the length of the median, first find the coordinates of midpoint M of $\overline{BC}$. If the midpoint is located at (x, y), then

$$x = \frac{-7 + 4}{2} \qquad \text{and} \qquad y = \frac{8 + 5}{2}$$

$$= \frac{-3}{2} \qquad\qquad\qquad = \frac{13}{2}$$

Thus the coordinates of midpoint M of $\overline{BC}$ are $(\frac{-3}{2}, \frac{13}{2})$.

The length of $\overline{AM}$ will then be the length of the median to $\overline{BC}$.

$$AM = \sqrt{(\tfrac{-3}{2} - 2)^2 + (3 - \tfrac{13}{2})^2}$$
$$= \sqrt{(\tfrac{-7}{2})^2 + (\tfrac{-7}{2})^2}$$
$$= \sqrt{\tfrac{49}{4} + \tfrac{49}{4}}$$
$$= \sqrt{\tfrac{98}{4}}$$
$$= \tfrac{7}{2}\sqrt{2}$$

Thus the length of the median from A to $\overline{BC}$ is $\frac{7}{2}\sqrt{2}$.   ◀

In Sections 4.1 and 4.2 we studied the Pythagorean Theorem and its converse. In this section we use the converse of the Pythagorean Theorem.

EXAMPLE 13   In the figure shown, triangle ABC has vertices at $(0, 0)$, $(3, 4)$, and $(8\frac{1}{3}, 0)$, respectively. Show that the triangle is a right triangle.

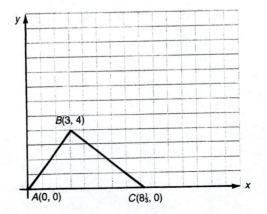

Solution   To show that the triangle is a right triangle we need to find the lengths of each side and then use the converse of the Pythagorean Theorem.

$$AB = \sqrt{(3-0)^2 + (4-0)^2}$$
$$= \sqrt{9 + 16}$$
$$= \sqrt{25}$$
$$= 5$$

$$BC = \sqrt{(3 - 8\frac{1}{3})^2 + (4-0)^2} \qquad\qquad AC = \sqrt{(8\frac{1}{3} - 0)^2 + (0-0)^2}$$
$$= \sqrt{(-5\frac{1}{3})^2 + 16} \qquad\qquad\qquad = \sqrt{(8\frac{1}{3})^2}$$
$$= \sqrt{(\frac{-16}{3})^2 + 16} \qquad\qquad\qquad = 8\frac{1}{3}$$
$$= \sqrt{\frac{256}{9} + 16} \qquad\qquad\qquad\quad = \frac{25}{3}$$
$$= \sqrt{\frac{256 + 144}{9}}$$
$$= \sqrt{\frac{400}{9}}$$
$$= \frac{20}{3}$$

Now we ask the following question: Is $(AB)^2 + (BC)^2 = (AC)^2$?

$$(5)^2 + (\tfrac{20}{3})^2 \overset{?}{=} (\tfrac{25}{3})^2$$
$$25 + \tfrac{400}{9} \overset{?}{=} \tfrac{625}{9}$$
$$\tfrac{625}{9} \overset{\checkmark}{=} \tfrac{625}{9}$$

Thus by the converse of the Pythagorean Theorem $\triangle ABC$ is a right triangle with right angle at B.   ◀

QUICK CHECK

1. A triangle has vertices located at $(0, 0)$, $(6, 0)$, and $(3, 3\sqrt{3})$. Show that the triangle is equilateral.

2. Triangle ABC has vertices at $(0, 0)$, $(3, 0)$, and $(0, 4)$, respectively. Find the length of the median to $\overline{BC}$.

3. Triangle PQR has vertices at $(2, -4)$, $(5, 1)$, and $(0, 4)$. Show that the triangle is a right triangle.

ANSWERS

1. Since each side has length 6, it is equilateral.
2. Midpoint M of $\overline{BC}$ is at $(\frac{3}{2}, 2)$. Thus the length of median AM is $\frac{5}{2}$.
3. Since $PQ = \sqrt{34}$, $QR = \sqrt{34}$, $PR = \sqrt{68}$, it follows that $(PQ)^2 = 34$, $(QR)^2 = 34$, and $(PR)^2 = 68$. Thus $(PQ)^2 + (QR)^2 = (PR)^2$. Therefore the triangle has a right angle at Q by the converse of the Pythagorean Theorem.

<u>A</u>

In Exercises 1–6, find the distance between each pair of points.

ANSWERS

1. $(-3, 6), (4, -5)$

2. $(-5, 4), (6, -3)$

3. $(-8, 5), (-8, 3)$

4. $(5, 6), (5, 9)$

5. $(4, 9), (-6, 9)$

6. $(3, -8), (5, -8)$

1. _____

2. _____

3. _____

4. _____

5. _____

6. _____

In Exercises 7–14, find the coordinates of the midpoints of the line segments whose endpoints are given.

7. $(-3, 5), (4, -6)$

8. $(-7, 2), (8, -3)$

9. $(4, 7), (-3, 7)$

10. $(2, -8), (5, -8)$

11. $(3, 4), (3, -7)$

12. $(6, 5), (6, -8)$

13. $(-2, 0), (7, 0)$

14. $(-8, 0), (-5, 0)$

7. _____

8. _____

9. _____

10. _____

11. _____

12. _____

13. _____

14. _____

In Exercises 15–17, the first given point indicates the coordinates of the midpoint of a line segment and the second given point indicates the coordinates of one endpoint of the line segment. In each instance find the coordinates of the other endpoint of the line segment.

15. $(4, 0), (-2, 7)$

16. $(-3, 5), (6, 0)$

17. $(5, 9), (-3, 2)$

18. Triangle *ABC* has vertices at $(-2, -1)$, $(-2, 4)$, and $(3, 4)$, respectively. Show that the triangle is an isosceles right triangle.

19. Triangle *PQR* has vertices at $(-2, 3)$, $(5, 3)$, and $(5, -4)$, respectively. Show that the triangle is an isosceles right triangle.

20. Triangle *RST* has vertices at $(-2, 7)$, $(4, 8)$, and $(-6, 5)$, respectively. Show that the triangle is scalene.

21. Triangle *UVT* has vertices at $(-5, 6)$, $(4, 8)$, and $(9, -9)$, respectively. Show that the triangle is scalene.

15. _____

16. _____

17. _____

18. _____

19. _____

20. _____

21. _____

B

22. The vertices of parallelogram $ABCD$ are $(-2, 2)$, $(-1, 6)$, $(6, 4)$, and $(5, 1)$. Find the point where the diagonals of the parallelogram intersect.

22. _____

23. The vertices of parallelogram $PQRS$ are $(-4, 2)$, $(1, 5)$, $(2, 0)$, and $(-3, -3)$. Find the point where the diagonals of the parallelogram intersect.

23. _____

ANSWERS

1. The following triangles are congruent as marked. Name a congruency correspondence between the triangles.

1. _____

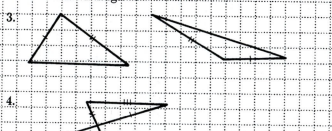

2. If △PQR ≅ △XYZ, name the side of △XYZ that $\overline{PR}$ is congruent to.

2. _____

In Exercises 3–6 decide whether or not the pairs of triangles are congruent based on the markings.

3.

3. _____

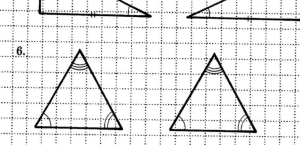

4.

4. _____

5.

5. _____

6.

6. _____

In Exercises 7–10 each of the pairs of triangles are congruent by one of the Rules for Congruent Triangles. Name the rule and list the congruent parts.

7.

7. _____

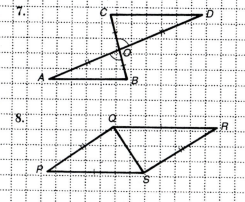

8.

8. _____

291

9.

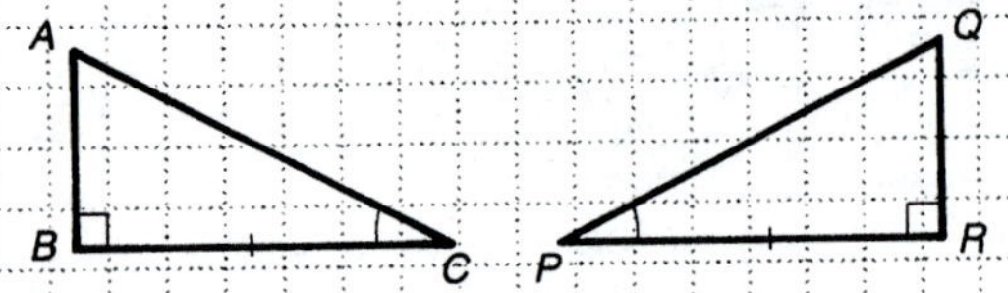

9. _____

10.

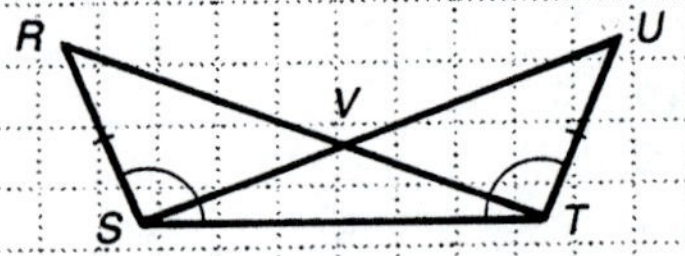

10. _____

11. In the figure shown triangles RST and UVW are similar. If $RS = \frac{1}{2}$, $UV = 2$, and $RT = \frac{3}{4}$, find UW.

11. _____

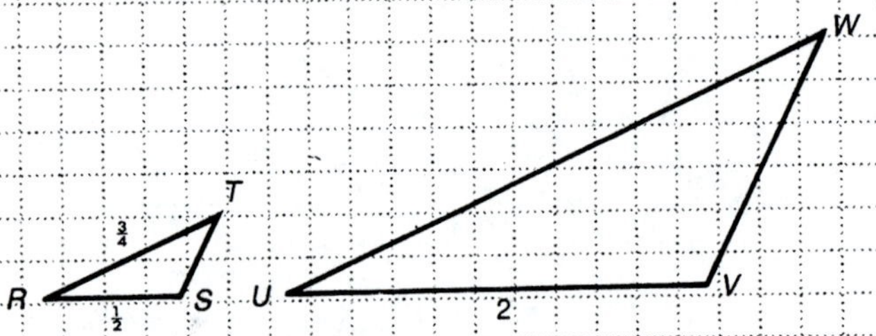

12. In the figure shown the triangles are similar as marked. Write a similarity correspondence for the triangles.

12. _____

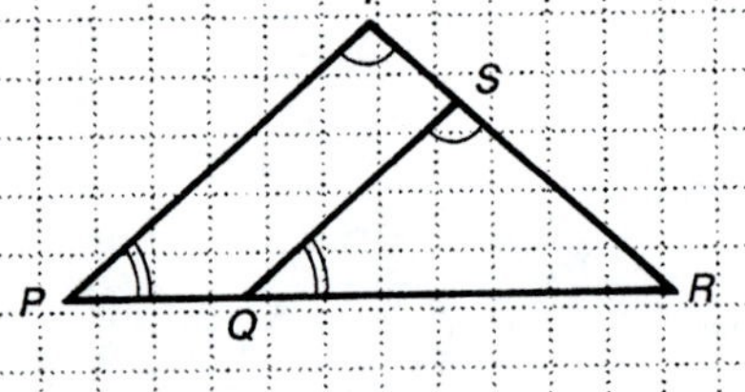

13. In the figure shown $\overline{AB} \parallel \overline{CD}$. Find two pairs of congruent angles in triangles AOB and DOC and write a similarity correspondence for the triangles.

13. _____

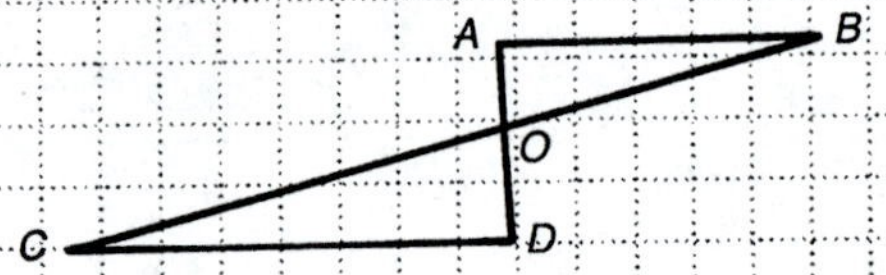

For Exercises 14–16 find the unknowns.

14.

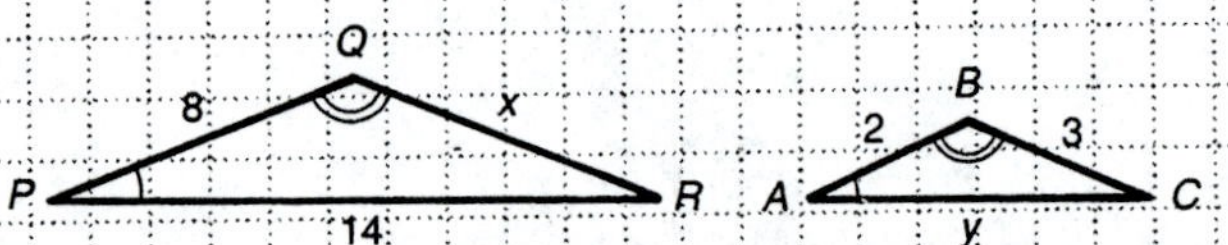

14. _____

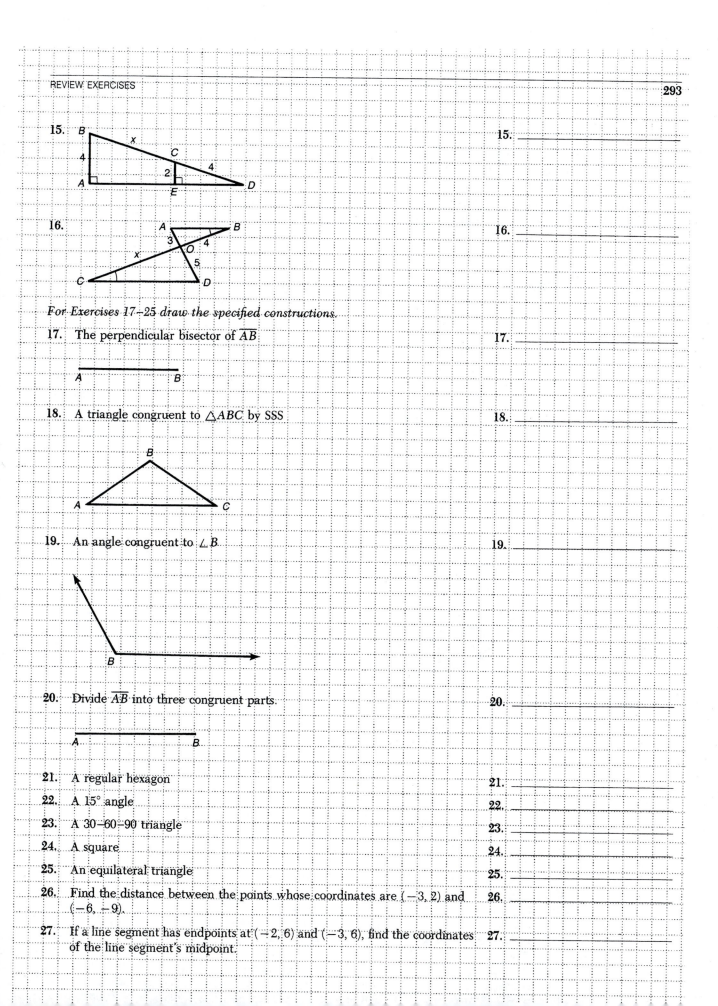

15.   15. _____

16.   16. _____

For Exercises 17–25 draw the specified constructions.

17.   The perpendicular bisector of $\overline{AB}$ 17. _____

18.   A triangle congruent to $\triangle ABC$ by SSS   18. _____

19.   An angle congruent to $\angle B$ 19. _____

20.   Divide $\overline{AB}$ into three congruent parts.   20. _____

21.   A regular hexagon   21. _____

22.   A 15° angle   22. _____

23.   A 30–60–90 triangle   23. _____

24.   A square   24. _____

25.   An equilateral triangle   25. _____

26.   Find the distance between the points whose coordinates are $(-3, 2)$ and   26. _____
      $(-6, -9)$.

27.   If a line segment has endpoints at $(-2, 6)$ and $(-3, 6)$, find the coordinates   27. _____
      of the line segment's midpoint.

28. The midpoint of a line segment is at $(-3, 9)$ and one endpoint is at $(0, 9)$. Find the coordinates of the other endpoint.

28. _____

29. The vertices of a triangle are $(-3, 0)$, $(1, 5)$, and $(5, 0)$. Is the triangle isosceles? Why or why not?

29. _____

30. A rhombus has vertices at $(0, 0)$, $(6, 8)$, $(10, 0)$, $(16, 8)$. Find the lengths of the diagonals.

30. _____

In Problems 1–3 the triangles are congruent as marked. Name a congruency correspondence for each pair of triangles.

ANSWERS

1.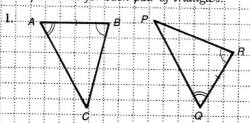

1. _____

2.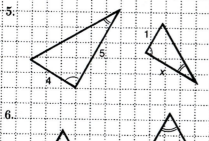

2. _____

3.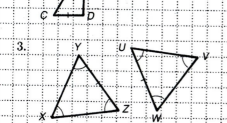

3. _____

4. If △ABC ≅ △PQR, list the corresponding congruent parts.

4. _____

In Problems 5–7 the triangles are similar as marked. Find the unknown.

5.

5. _____

6.

6. _____

7.

7. _____

For Problems 8–15 draw the specified constructions.

8. A line segment congruent to $\overline{PQ}$

8. _____

P •————————• Q

9. An angle congruent to $\angle A$

9. _____

10. The bisector of $\overline{PQ}$

10. _____

11. The bisector of $\angle R$

11. _____

12. A triangle congruent to $\triangle ABC$ by ASA

12. _____

13. A circle containing points A, B, and C

13. _____

14. A regular hexagon

14. _____

15. A 30° angle

15. _____

16. Find the distance between two points whose coordinates are $(-2, 9)$ and $(3, -8)$.

16. _____

17. Find the midpoint of the line segment whose endpoints are located at $(-3, 7)$ and $(4, 5)$.

17. _____

18. The midpoint of a line segment is located at $(3, 4)$ and one endpoint is at $(-3, 1)$. Find the coordinates of the other endpoint.

18. _____

19. The vertices of a parallelogram are located at $(0, 0)$, $(2, 5)$, $(6, 0)$, and $(8, 5)$. Find the length of the diagonals.

19. _____

20. The vertices of a triangle are located at $(0, 3)$, $(4, -8)$, and $(2, 7)$. Is the triangle isosceles, equilateral, or scalene?

20. _____

EXERCISE 1.1

1. False **3.** True **5.** True **7.** Not a mathematical definition because it does not tell how an orange can be distinguished from other fruits. **9.** Not a mathematical definition because it does not tell how a horse can be distinguished from other four-legged animals. **11.** Is a mathematical definition **13.** Is a mathematical definition **15.** Is a mathematical definition

EXERCISE 1.2

1. False **3.** True **5.** False **7.** False **9.** If it is a robin, then it is a bird. **11.** If I make a 300-mile car trip, then I will be tired. **13.** If I eat strawberries, then I will have hives. **15.** If it is hot, then the temperature is over 90 degrees. **17.** If it comes to a point, then it is a pyramid. **19.** It is a tree iff it is a plant. **21.** A number is a digit iff it is 0, 1, 2, 3, 4, 5, 6, 7, 8, or 9. **23.** A person is an Eskimo iff the person lives at the North Pole. **25.** A number is divisible by 11 iff it is a multiple of 11. **27.** A ball is a football iff it is the official ball used in the game of football.

EXERCISE 1.3

1. True **3.** False **5.** True **7.** True **9.** False **11.** True **13.** False **15.** True

EXERCISE 1.4

1. b **3.** a **5.** d **7.** b **9.** b **11.** b **13.** d

EXERCISE 1.5

1. The elements of the set are determined without question. **3.** ϕ; $\{1, 2, 3\}$; $\{1, 2\}$; $\{1, 3\}$; $\{2, 3\}$; $\{1\}$; $\{2\}$; $\{3\}$ **5.** If $A = \{2, 3\}$ and $B = \{1, 2, 3\}$, then $A \subseteq B$. If $R = \{$all people in the United States$\}$ and $S = \{$women in the United States$\}$, then $R \subseteq S$. **7.** If $A = \{1, 2, 3\}$ and $B = \{2, 1, 3\}$, then $A = B$. If $P = \{$the digits less than 5$\}$ and $Q = \{0, 1, 2, 3, 4\}$, then $P = Q$. **9.** $\{1, 2, 3, 4\}$; $\{1, 3, 5, 7, 9, 11, \ldots, 19\}$ **11.** Roster **13.** Roster **15.** Rule **17.** $\{0, 2, 3, 4, 5, 6, 7, \ldots\}$ **19.** $\{$the two most popular domestic pets$\}$ **21.** False **23.** True **25.** False **27.** True **29.** True **31.** False **33.** False **35.** True **37.** False **39.** True **41.** False **43.** True **45.** True **47.** True **49.** True **51.** False **53.** True

EXERCISE 1.6

1. $\{1, 2, 3\} \cap \{2, 3, 4\} = \{2, 3\}$; $\{1, 2, 3\} \cap \{4, 5, 6\} = \phi$ Answers may vary. **3.** $\{1, 2, 3\} \cup \{2, 3, 4\} = \{1, 2, 3, 4\}$; $\{1, 2, 3\} \cup \{4, 5, 6\} = \{1, 2, 3, 4, 5, 6\}$ Answers may vary. **5.** A **7.** Yes **9.** No **11.** $\{5, 6, 7\}$ **13.** ϕ **15.** $\{2, 3, 4, 5, 6, 7, 8, 9, 10, 11, 12, 13, 14, 15\}$ **17.** U **19.** $\{2, 3, 4, 5, 6, 7, 8, 9, 10, 11, 14, 15\}$ **21.** 50 **23.** 100
25.

27. 4 **29.** 11

CHAPTER 1 REVIEW EXERCISES

1. It does not distinguish wrens from other birds. **2.** The converse is false. **3.** It does not distinguish circles from other round objects. **4.** False **5.** False **6.** True **7.** True **8.** True **9.** False **10.** False **11.** True **12.** True **13.** False **14.** False **15.** False **16.** True **17.** True **18.** True **19.** False **20.** False **21.** True **22.** False **23.** False **24.** True **25.** False **26.** True **27.** True **28.** $\in$ **29.** $\notin$ **30.** $\{8\}$ **31.** $A \subset B$ **32.** $Z \not\subseteq R$ **33.** $\cup$ **34.** $\cap$ **35.** $P \cap Q$ **36.** A method for describing sets in which the elements are all listed or a partial list of elements establishes a pattern from which all elements may be determined. **37.** A method for describing sets in which the elements are described using words. **38.** Sets having exactly the same elements. **39.** Sets having no elements in common. **40.** False **41.** False, because they have elements in common. **42.** False, because the union of A and B contains 0, 1, 2, 3, . . . , 10, not just 1, 2, and 3. **43.** False, because $A \cap B = A$. **44.** True **45.** True **46.** False, because the elements 1 and 3 of set F are not in set D. **47.** False, because $A \cap C = \{1, 2, 3, \ldots, 9\}$

48.   **49.** {1, 2, 3, 4, 5}   **50.** *E, G*   **51.** No   **52.** *A*   **53.** *F*   **54.** *C*   **55.** Yes   **56.** No

57. *B* and *D*; *B* and *G*   **58.** *E*   **59.** No   **60.** Yes   **61.** 2   **62.** 23   **63.**

CHAPTER 1 PRACTICE TEST

1. Terms you are expected to know intuitively.
2. A mathematical definition must place the word being defined into a set and then show how it can be distinguished from all other members of the set.
3. Properties are fundamental statements that are accepted without justification. Rules can be justified based on properties, definitions, and other previously proven rules.
4. Biconditional statement
5. A conditional statement is written in the if–then form. A biconditional statement is a conditional statement and its converse joined by the word and.
6. If it is a dog, then it is a domestic animal.   **7.** If a number is divisible by 2, then it is even.
8. A number is divisible by 3 iff it is a multiple of 3.   **9.** It is a digit iff it is 0, 1, 2, 3, 4, 5, 6, 7, 8, or 9.   **10.** They are both true.
11. Weslie will buy a white car, a four-door car, or a white four-door car.   **12.** Carla went to both San Antonio and to Houston.
13. c   **14.** b   **15.** c   **16.** a   **17.** a   **18.**   **19.** *B*   **20.** {5, 6, 7}   **21.** {6, 7, 8}   **22.** *D*   **23.** *B*

24. *A*   **25.** {1, 2, 3, 4, 5, 6, 7, 9, 10, 12, 13}   **26.** *U*   **27.** No   **28.** Yes   **29.** Yes   **30.** Yes   **31.** 24   **32.** 66   **33.** ↑

EXERCISE 2.1

1. 3520   **3.** 0.057   **5.** 253,440   **7.** 72   **9.** 31.$\overline{56}$ **11.** 6   **13.** 3.5   **15.** 7,000,000   **17.** 5000   **19.** 0.65   **21.** 0.192
23. 4.04   **25.** 18.3   **27.** 19.05   **29.** 0.763   **31.** 8.05 kℓ   **33.** 128.7 hm   **35.** 162.56 cm   **37.** 27.4 m   **39.** 60.96 m   **41.** 19 in.
43. 115 mi   **45.** 3.1 mi   **47.** 196,850 ft   **49.** 708.66 in.   **51.** 5.5 mi   **53.** 6.77 mi   **55.** 8576 ft   **57.** 3 in.

EXERCISE 2.2

1. 0.1   **3.** 4096   **5.** 169.6   **7.** 800   **9.** 160.1875   **11.** 40   **13.** 1500   **15.** 20,000   **17.** 500   **19.** 2,000,000   **21.** 50
23. 572   **25.** 738,636   **27.** 1136   **29.** 0.909   **31.** 10.45 kg   **33.** 1.36 metric tons   **35.** 0.005 oz   **37.** 33 lb   **39.** 0.066 lb
41. 1.136 metric tons   **43.** No   **45.** $1.53   **47.** $6.36/kg

EXERCISE 2.3

1. 7   **3.** 20   **5.** 10   **7.** 12   **9.** 2.5   **11.** 40,000   **13.** 0.096   **15.** 2,200,000   **17.** 300   **19.** 85,000   **21.** 2.12   **23.** 0.477
25. 53   **27.** 221.7   **29.** 203.8   **31.** 3.18 qt   **33.** 7.5 ℓ   **35.** 5.9 dℓ   **37.** 15.1 ℓ   **39.** 15,000,000 mm   **41.** $0.30   **43.** $1.97
45. $47.00/ℓ

EXERCISE 2.4

1. $232\frac{2}{9}$ **3.** $4\frac{4}{9}$ **5.** $12\frac{7}{9}$ **7.** $43\frac{1}{3}$ **9.** $48\frac{8}{9}$ **11.** 140   **13.** 194   **15.** 32   **17.** 231.8   **19.** 267.8   **21.** 37°C   **23.** $25\frac{5}{9}$°C
25. $4\frac{4}{9}$°C   **27.** 842°F   **29.** 167°F   **31.** 5538°C   **33.** 29.1°C   **35.** 61.1°C   **37.** $268.16

EXERCISE 2.5

1. 2   **3.** 12   **5.** 16   **7.** 24   **9.** 64   **11.** 1000   **13.** 2500   **15.** 200   **17.** 150.8   **19.** 900,000   **21.** $3\frac{3}{4}$ **23.** $4\frac{11}{16}$
25. $7\frac{13}{16}$ **27.** 0.6   **29.** 0.273⁻   **31.** 217.65   **33.** 442.2   **35.** 8.84   **37.** 1.187   **39.** 678.4   **41.** $0.64   **43.** $1.28
45. $0.026/fl oz

CHAPTER 2 REVIEW EXERCISES

1. 20 **2.** 1440 **3.** $6\frac{1}{2}$ **4.** 15,840 **5.** 0.15 **6.** 20 **7.** 3 **8.** 5 **9.** 58.42 **10.** 71.08 **11.** 40.4 **12.** 2,952,756
13. 640 **14.** 32,000 **15.** 1400 **16.** 0.0019 **17.** 30.8 **18.** 73,864 **19.** 18.2 **20.** 40 **21.** $1\frac{1}{2}$ **22.** 0.08 **23.** 0.14
24. 8.5 **25.** 0.4 **26.** 98.1 **27.** 66 **28.** 90.1 **29.** 0.6 **30.** 29.4 **31.** 2.22 **32.** 104 **33.** 185 **34.** 120 **35.** 16
36. 96 **37.** 42 **38.** 24

CHAPTER 2 PRACTICE TEST

1. $14\frac{1}{3}$ **2.** $6\frac{7}{12}$ **3.** 4.72 **4.** 68.9 **5.** 0.523 **6.** 112.7 **7.** 2,000,000 **8.** 192 **9.** 35.2 **10.** 2900 **11.** 10.9 **12.** 4.25
13. 52.8 **14.** 2.5 **15.** 1.7 **16.** 168.8 **17.** $2\frac{7}{8}$ **18.** 96

EXERCISE 3.1

1. False. To get a plane you must have three points. **3.** True **5.** False, because their intersection can be a plane or a line.
7. False, because $\overrightarrow{AB}$ and $\overrightarrow{BA}$ point in different directions, so they cannot be the same set of points. **9.** True
11. False. Between cannot be length. **13.** False, because they can be skew lines. **15.** True
17. False, because three noncollinear points determine a plane. **19.** True
21. Intersecting planes, $\mathcal{N}$ and $\mathcal{M}$; perpendicular planes, $\mathcal{N}$ and $\mathcal{P}$; parallel planes, $\mathcal{O}$ and $\mathcal{P}$ **23.** $\overline{AC}$ **25.** $\overrightarrow{AB}$ **27.** $\{B\}$ **29.** $\overrightarrow{BE}$
31.   **33.** 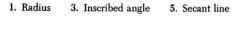 **35.**   **37.**   **39.** Four points need not be in the same plane.

41. 10

EXERCISE 3.2

1. $\angle ABC$, $\angle CBA$, $\angle B$ **3.** $\overrightarrow{BA}$, $\overrightarrow{BC}$ **5.** Q is in the exterior of the angle. **7.** 36° **9.** 131° 39′ 58″ **11.** 10° 55′ 37″
13. 63° 16′ 5″ **15.** 257° 33′ **17.** True **19.** True **21.** False **23.** True **25.** True **27.** True **29.** True **31.** True
33. False **35.** $\angle R$ **37.** B is in the interior of the triangle. **39.** $\triangle PQR$ **41.** 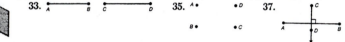 **43.**
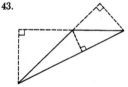

45. 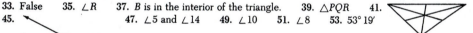 **47.** $\angle 5$ and $\angle 14$ **49.** $\angle 10$ **51.** $\angle 8$ **53.** 53° 19′

55. 74° 58′ 6″ **57.** 61° 20′ and 118° 40′

EXERCISE 3.3

1. e **3.** b, c, d, f **5.** n **7.** b, f **9.** a **11.** l **13.** e, n, o **15.** p **17.** g, h **19.** None **21.** j **23.** a, b, c, e
25. 150° **27.** 48°, 88°, 44° **29.** False **31.** False **33.** True **35.** True **37.** True **39.** $x = 70° 24′$ **41.** $x = 8$
43. $x = 60°$ **45.** m $\angle 1 +$ m $\angle 2 +$ m $\angle 3 = 180°$; m $\angle 1 =$ m $\angle 5$; m $\angle 3 =$ m $\angle 4$; m $\angle 5 +$ m $\angle 2 +$ m $\angle 4 = 180°$

EXERCISE 3.4

1. Radius **3.** Inscribed angle **5.** Secant line **7.** **9.** **11.** **13.** False **15.** False

17. False **19.** False **21.** True **23.** $x = 100°$ **25.** $x = 10°$ **27.** $x = 55°$ **29.** $x = 60°$ **31.** $x = 45°$ **33.** $x = 40°$
35. 40° **37.** 125°

EXERCISE 3.5

1. False **3.** False **5.** False **7.** False **9.** True **11.**   **13.**   **15.**   **17.**

19.    21. 7   23. 3   25. 8   27.   29.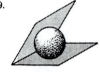

CHAPTER 3 REVIEW EXERCISES

1. Triangle *ABC*   2. Arc *PDQ*   3. Ray *AB*   4. The measure of line segment *BD*   5. The measure of arc *RST*   6. Line *AZ*
7. The interior of line segment *VT*   8. Half-line *RK*   9. Angle *A*   10. The measure of angle *PTZ*   11. Line   12. Skew
13. Ray   14. Dodecagonal   15. Collinear   16. Space   17. Line   18. Plane   19. Coplanar, collinear   20. Parallel
21. Bisect   22. Opposite   23. Angle   24. Three angles   25. Same measure   26. Acute angle   27. 90   28. Straight angle
29. Obtuse angle   30. Perpendicular   31. Complementary   32. 180°   33. Isosceles   34. No   35. 60°   36. One right angle
37. Quadrilateral   38. Equilateral   39. 180°   40. 540°   41. Parallelogram   42. Trapezoid   43. Rhombus   44. Square
45. Supplementary   46. Congruent   47. Perpendicular   48. Octagon   49. Diameter   50. Secant   51. Concentric
52. Tangent   53. Intersects, one point   54. Central angle   55. Sphere   56. Parallelogram   57. Circle   58. Polygon
59. Polygons   60. Circles   61. Twice   62. Parallelepiped   63. Parallelepiped   64. Dodecagon   65. Central angle
66. Quadrilateral   67. Isosceles   68. Polygonal   69. Parallel   70. Diameter   71. Pyramid   72. Inscribed   73. Vertex
74. Arc   75. Supplementary   76. $\angle A$ and $\angle B$ 77. $\overline{AB}$ and $\overline{BC}$ 78. $\overline{AC}$ 79. $\overline{AB}$ and $\overline{DC}$ 80. $\angle A$ and $\angle C$ 81. *a* and *e*
82. *b* and *f*   83. *a* and *g*   84. *b* and *e*   85. 89° 45′ 14″   86. 24° 58′ 9″   87. 7° 42′ 18″   88. 75°   89. 160°
90. 62° 20′, 70° 20′, 47° 20′   91. 144°   92. $7\frac{1}{2}$ inches   93. 62°   94. 70°   95. 45°   96. 20°   97. c   98. f   99. d
100. e   101. a   102. b   103. g   104. h

CHAPTER 3 PRACTICE TEST

1. Line segments having the same measure   2. The union of two rays with a common endpoint
3. A point that divides a line segment into two congruent line segments   4. The union of a half-line with its endpoint
5. Two lines that intersect forming right angles   6. An angle whose measure is greater than 0° and less than 90°
7. A quadrilateral with two pairs of parallel sides   8. A quadrilateral with exactly one pair of parallel sides
9. The set of points in a plane a given distance from a given point.   10. A chord of a sphere that contains the center of the sphere
11. Angle *A*   12. The measure of angle *A*   13. The measure of line segment *AB*   14. The measure of arc *PQR*
15. Line *AB* is perpendicular to line *PQ*   16. Line *l* is parallel to line *n*   17. Triangle *PQR*   18. Ray *RS*
19. The interior of line segment *AB*   20. Half-line *RT*   21. By two distinct points   22. By the intersection of two lines
23. By three noncollinear points   24. By four noncoplanar points   25. By three noncollinear points
26. They have the same size and shape.   27. $x = 50°$ 28. $x = 30°$ 29. $x = 20°$ 30. $x = 20°$ 31. $x = 20°$ 32. $x = 70°$
33. $x = 47° 30′$ 34. $x = 60°$ 35. a   36. d   37. q   38. m   39. b   40. f   41. o   42. i   43. c   44. p   45. h
46. g   47. p   48. n   49. j   50. k   51. l   52. e

EXERCISE 4.1

1. $x = 10$ 3. $x = \sqrt{337}$ 5. $x = 12$ 7. $x = 24$ 9. $x = \sqrt{3}$ 11. 10   13. $2\sqrt{3}$ 15. $3\sqrt{5}$ 17. $\sqrt{285}$ 19. $\sqrt{481}$
21. $2\sqrt{73}$ 23. 40 ft   25. 2 miles shorter   27. $6\sqrt{3}$ 29. $2\sqrt{41}$ 31. $\sqrt{65}$ 33. 14 in.   35. $2\sqrt{6}$

EXERCISE 4.2

1. $x = \dfrac{4}{\sqrt{3}}$ cm, $y = \dfrac{8}{\sqrt{3}}$ cm   3. $x = 16$ dm, $y = 16\sqrt{3}$ dm   5. $x = 5$ in.   7. $x = 8$ m, $y = 4$ m   9. $x = 3$ dm   11. $x = 5\sqrt{6}$ hm
13. $x = \frac{2}{2}\sqrt{10}$ mi   15. $x = 3\sqrt{2}$ dm, $y = 3\sqrt{6}$ dm   17. $x = 7$ m, $y = 14$ m, $z = 7\sqrt{3}$ m   19. $x = 3$ ft   21. $x = 8$ cm   23. $x = 4$ in.
25. $x = 12$ ft   27. $\frac{5}{2}\sqrt{3}$ in.   29. $5\sqrt{2}$ in.

EXERCISE 4.3

1. 16 ft   3. 36 cm   5. 52 in.   7. 18π cm   9. 50 ft   11. 144 ft   13. 30 ft   15. 30π in.   17. $34 + 34\sqrt{2}$ ft   19. 46 in.
21. 40 ft   23. $\dfrac{8}{\pi}$ in.   25. $\dfrac{39}{2\pi}$ in.   27. 7 ft   29. $\dfrac{\sqrt{2581}}{2}$ in.   31. 24 in.   33. 32 ft

CHAPTER 4 REVIEW EXERCISES

1. $\sqrt{74}$ 2. $\sqrt{161}$ in.   3. $x = 8$, $y = 4\sqrt{3}$ 4. $x = 3\frac{1}{2}$, $y = 3\frac{1}{2}\sqrt{3}$ 5. $y = \dfrac{17}{\sqrt{3}}$, $x = \dfrac{34}{\sqrt{3}}$ 6. $x = 72\sqrt{2}$ 7. $x = \dfrac{43}{\sqrt{2}}$ 8. $x = 10$
9. $x = 26$ 10. $x = 2\sqrt{55}$ 11. $x = 4\frac{1}{2}\sqrt{3}$ 12. 84 in.   13. 112 in.   14. $228 + 76\sqrt{3}$ ft   15. 68π dm, 267.7π in.   16. 13.7 cm

CHAPTER 4 PRACTICE TEST

1. $x = 4\sqrt{3}$ m, $y = 8$ m   2. $x = \dfrac{15}{\sqrt{3}}$ dm, $y = \dfrac{30}{\sqrt{3}}$ dm   3. $x = 7\sqrt{2}$ cm   4. $\sqrt{73}$ in.   5. 10 ft   6. 30 in.   7. 24   8. 40   9. 28
10. 30   11. 4π in.   12. 3π in.

EXERCISE 5.1

1. 35 ft²   3. 120 dm²   5. 9 in.²   7. 128 ft²   9. 72 m²   11. $20.25\sqrt{3}$ m²   13. 16 yd²   15. 24.5 in.²   17. $(6 + 16\sqrt{2})$ m²
19. 48 cm²   21. 36 yd²   23. $300\sqrt{3}$ yd²   25. $288\sqrt{3}$ cm²   27. 54 mm²   29. $(37.5 + 57.5\sqrt{3})$ ft²   31. $85\sqrt{3}$ ft²
33. $73.5\sqrt{3}$ mm²   35. $512\sqrt{3}$ km²   37. 47.61π mm²   39. 175.17π m²   41. 5 m   43. 1.2 mm   45. $\frac{13}{72}$ ft²   47. 36 ft²
49. $3\frac{5}{9}$ yd²   51. 5760 in.²   53. 3,097,600 yd²   55. 2500 dm²   57. 360,000 m²   59. 20 cm²   61. 0.0003 dam²
63. 2,000,000 mm²   65. 10.03 m²   67. 678.2 dm²   69. 14.84 cm²   71. 1.12 mi²   73. 495.14 ft²   75. 72,600 yd²   77. 14.82 ac
79. 54.34 ac   81. 1.21 ha   83. $\frac{1}{4}$ ft²   85. 0.09 m²   87. 1.35 ft²   89. 17,280 in.²   91. 18.6 in.²   93. 10.12 ha   95. 10.3 ac
97. 1920 ac   99. 400,000 sections   101. 20.42, but they will probably have to buy $20\frac{1}{2}$ square yards at a cost of $266.50   103. 3717.12 lb
105. $8\pi + 16\sqrt{3}$ in.²   107. $(4\sqrt{3} - 2\pi)$ in.²

EXERCISE 5.2

1. 144π in.²   3. 90π ft²   5. 96π cm²   7. 100π cm²   9. 1350 m²   11. $64 + 16\sqrt{33}$ ft²   13. $336 + 192\sqrt{3}$ in.²
15. $(42 + \frac{21}{2}\sqrt{39} + 6\sqrt{91})$ yd²   17. 479.90 in.²   19. $(240 + 75\sqrt{3})$ in.²   21. $16\sqrt{3}$ m²   23. $(4128 + 512\sqrt{3})$ cm²   25. 1800 ft²
27. 4.72 gal   29. $2126\frac{1}{4}\pi$ ft²

EXERCISE 5.3

1. 0.023 ft³   3. $33\frac{1}{3}$ yd³   5. 12,501 ft³   7. 454,464 in.³   9. 0.000092 hm³   11. 0.2 dm³   13. 0.000000525 dm³   15. 2 cm³
17. 819.35 cm³   19. 651.3 dm³   21. 54.9 yd³   23. 1.2 mi³   25. $\frac{256}{3}$ m³   27. $11\frac{2}{3}\pi$ in.³   29. 96π cm³   31. 288 dam³
33. $367.5\sqrt{3}$ ft³   35. 367.5 ft³   37. 1251.4 in.³   39. 80 hm³   41. $2850\sqrt{3}$ mm³   43. 288 ft³   45. $216\sqrt{3}$ ft³   47. $272\sqrt{5}$ ft³
49. $\frac{4913}{12}\sqrt{2}\pi$ mm³   51. 102π in.³   53. $\frac{8}{3}\pi$ in.³

CHAPTER 5 REVIEW EXERCISES

1. $112\frac{1}{2}$ cm²   2. 112 m²   3. 20 in.²   4. 180 hm²   5. 90 ft²   6. 1382.15 in.²   7. $\dfrac{96}{\sqrt{3}}$ cm²   8. π m²   9. $4500\sqrt{41}\pi$ cm²

10. 4.6 cm   11. 367.5π 12. 523.7 dm²   13. 652 m²   14. 24,416.6 yd³   15. 800π cm³   16. 9.08 m³   17. $84\sqrt{3}$ in.³
18. 125 cm³   19. 3240 ft³   20. $\frac{1}{36}$ ft²   21. $2\frac{7}{9}$ 22. 0.03 m²   23. 2.97 m³   24. $1\frac{8}{27}$ yd³   25. 3,905,519.6 in.³

CHAPTER 5 PRACTICE TEST

1. 64π ft   2. 4π cm²   3. 24 ft²   4. $8\sqrt{3}$ in.²   5. 243 ft²   6. 192 in.²   7. 96 mm²   8. 4500π cm³   9. 100π ft³   10. 320 ft³
11. 1372π cm³   12. 144π ft²   13. 517.7 m²   14. $\frac{83}{144}$ ft²   15. 16.896 in.²   16. 465.6 yd³   17. 6.11×10^{-9}

EXERCISE 6.1

1. $\triangle ABC \cong \triangle STR$ 3. $\triangle PTV \cong \triangle YXW$ 5. $\triangle ROZ \cong \triangle TSV$ 7. $\angle R \cong \angle U$, $\angle S \cong \angle V$, $\angle T \cong \angle W$, $\overline{RS} \cong \overline{UV}$, $\overline{ST} \cong \overline{VW}$, $\overline{RT} \cong \overline{UW}$
9. $\angle PQR \cong \angle SRQ$, $\angle RPQ \cong \angle QSR$, $\angle QRP \cong \angle RQS$, $\overline{PQ} \cong \overline{SR}$, $\overline{QR} \cong \overline{RQ}$, $\overline{PR} \cong \overline{SQ}$
11. $\angle A \cong \angle R$, $\angle B \cong \angle S$, $\angle C \cong \angle T$, $\overline{AB} \cong \overline{RS}$, $\overline{BC} \cong \overline{ST}$, $\overline{AC} \cong \overline{RT}$ 13. SAS; $\overline{RW} \cong \overline{KL}$, $\angle W \cong \angle L$, $\overline{WV} \cong \overline{LJ}$
15. ASA; $\angle B \cong \angle C$, $\overline{BO} \cong \overline{OC}$, $\angle AOB \cong \angle DOC$ 17. ASA; $\angle P \cong \angle N$, $\overline{PA} \cong \overline{NB}$, $\angle A \cong \angle B$ 19. Isosceles
21. Opposite sides of a parallelogram are congruent, so the two triangles are congruent by SSS.
23. All four sides of a rhombus are congruent, and the diagonals of a rhombus bisect each other. Thus, congruency of the four triangles can be shown by SSS.
25. If M is the point of intersection of $\overline{AB}$ and its perpendicular bisector, then $\triangle AMP \cong \triangle BMP$ by SAS. Thus $AB = BP$.

EXERCISE 6.2

1. $\triangle ABC \sim \triangle PRQ$ 3. $\triangle ACD \sim \triangle FEB$ 5. $\triangle RZN \sim \triangle MPW$ 7. $PR = 16$ 9. $BP = 2\frac{2}{5}$ 11. $WY = 6.4$
13. $\angle B \cong \angle C$ and $\angle A \cong \angle D$ 15. $\triangle ABC \sim \triangle ACD$, $\triangle ABC \sim \triangle CBD$, $\triangle ACD \sim \triangle CBD$ 17. $w = 51$, $z = 14$, $x = 40°$, $y = 50°$
19. $x = 3\frac{1}{2}$, $y = 3\frac{1}{2}\sqrt{3}$, $z = 2$, $w = 2\sqrt{3}$ 21. $x = \frac{922}{165}$, $y = \frac{1155}{384}$ 23. $x = 2\frac{2}{5}$, $y = 4\frac{1}{5}$ 25. $p = 60°$, $x = 4$ 27. True   29. False

EXERCISE 6.3

1.

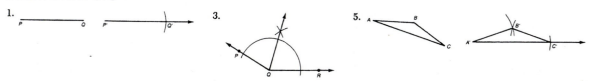

7. 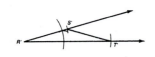   **9.**   **11.**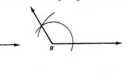

13. 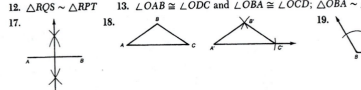   **15.** $\triangle ABC$ is equilateral. Then $\angle BAC$ is bisected, which makes m $\angle DAC = 30°$. Then $\angle DAC$ is bisected. Thus m $\angle EAC = 15°$.

17.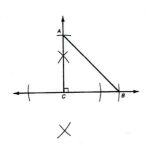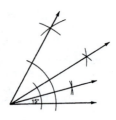

EXERCISE 6.4

1. $\sqrt{170}$ **3.** 2   **5.** 10   **7.** $(\frac{1}{2}, -\frac{1}{2})$ **9.** $(\frac{1}{2}, 7)$ **11.** $(3, -\frac{3}{2})$ **13.** $(\frac{5}{2}, 0)$ **15.** $(10, -7)$ **17.** $(13, 16)$
19. $\overline{PQ} = 7$, $\overline{QR} = 7$, and $\overline{PR} = 7\sqrt{2}$. Since $7^2 + 7^2 = (7\sqrt{2})^2$, $49 + 49 = 49.2$, and $98 = 98$, the triangle is a right triangle by the converse of the Pythagorean Theorem.   **21.** $\overline{UV} = \sqrt{85}$, $\overline{VT} = \sqrt{314}$, and $\overline{UT} = \sqrt{421}$ **23.** $(-1, 1)$

CHAPTER 6 REVIEW EXERCISES

1. $\triangle ABC \cong \triangle QRP$ **2.** $\overline{XZ}$ **3.** Not congruent   **4.** Congruent   **5.** Not congruent   **6.** Not congruent
7. SAS; $\overline{AO} \cong \overline{DO}$, $\angle AOB \cong \angle DOC$, $\overline{OB} \cong \overline{OC}$ **8.** SSS; $\overline{PQ} \cong \overline{RS}$, $\overline{PS} \cong \overline{RQ}$, $\overline{QS} \cong \overline{SQ}$ **9.** ASA; $\angle B \cong \angle R$, $\overline{BC} \cong \overline{RP}$, $\angle C \cong \angle P$
10. SAS; $\overline{RS} \cong \overline{UT}$, $\angle RST \cong \angle UTS$, $\overline{ST} \cong \overline{TS}$ **11.** $\overline{UW} = 3$
12. $\triangle RQS \sim \triangle RPT$ **13.** $\angle OAB \cong \angle ODC$ and $\angle OBA \cong \angle OCD$; $\triangle OBA \sim \triangle OCD$ **14.** $x = 12$, $y = \frac{7}{2}$ **15.** $x = 4$ **16.** $x = \frac{20}{3}$
17.   **18.**   **19.**

20. 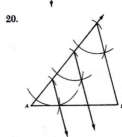   **21.** 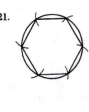   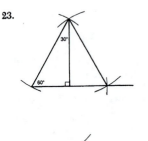   **23.**

24. 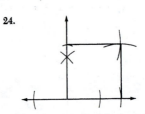   **25.**   **26.** $\sqrt{130}$ **27.** $(-\frac{5}{2}, 6)$ **28.** $(-6, 9)$

29. Yes, because the sides have lengths of $\sqrt{41}$, 8, and $\sqrt{41}$.   **30.** $8\sqrt{5}$ and $4\sqrt{5}$

CHAPTER 6 PRACTICE TEST

1. $\triangle ABC \cong \triangle QRP$　**2.** $\triangle AOB \cong \triangle DOC$　**3.** $\triangle XYZ \cong \triangle VWU$　**4.** $\angle A \cong \angle P, \angle B \cong \angle Q, \angle C \cong \angle R, \overline{AB} \cong \overline{PQ}, \overline{AC} \cong \overline{PR}, \overline{BC} \cong \overline{QR}$

5. $x = \frac{5}{4}$　**6.** $z = \frac{20}{3}$　**7.** $x = \frac{27}{5}$　**8.**

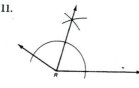

9.

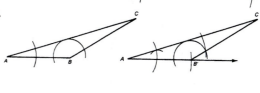

10.

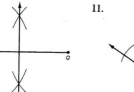

11.

12.

13.

14.

15.

16. $\sqrt{314}$　**17.** $(\frac{1}{2}, 6)$　**18.** $(9, 7)$　**19.** $\sqrt{41}, \sqrt{89}$

20. Scalene